D0770141

When the Lights Went Out

When the Lights Went Out

A History of Blackouts in America

David E. Nye

The MIT Press
Cambridge, Massachusetts
London, England

For information on special quantity discounts, email special_sales@ mitpress.mit.edu.

Set in Stone Sans and Stone Serif by Toppan Best-set Premedia Limited. Printed and bound in the United States of America.

Library of Congress Cataloging-in-Publication Data

Nye, David E., 1946–
When the lights went out: a history of blackouts in America/ David E. Nye.
    p. cm.
Includes bibliographical references and index.
ISBN 978-0-262-01374-1 (hbk. : alk. paper)
1. Electric power failures—United States—History. 2. Electrification—United States—History. 3. Electrification—Social aspects—United States. I. Title.
HD9685.U5N944   2010
333.793'20973—dc22

                                                        2009025494

10 9 8 7 6 5 4 3 2 1

dedicated to the Memory of Fern Drumheller Nye (May 5, 1920–March 18, 2008)

Contents

Acknowledgements   ix

Introduction   1

1 │ **Grid**   9

2 │ **War**   37

3 │ **Accident**   67

4 │ **Crisis**   105

5 │ **Rolling Blackouts**   137

6 │ **Terror**   173

7 │ **Greenout?**   205

Notes   233
Bibliography   269
Index   285

Acknowledgements

Every book is assisted by the generosity, interest, and support of others, and however valuable the Internet, libraries remain indispensable to research. I worked at the Danish Royal Library in Copenhagen, several libraries at the Massachusetts Institute of Technology, the Boston Public Library, the Rothermere Library at Oxford University, Georgetown University's library, the Robert Frost Library at Amherst College, the W. E. B. Dubois Library at the University of Massachusetts at Amherst, and the library at the University of Southern Denmark. I also thank the Edison Electric Research Institute in Washington for allowing me access to its collections. In addition, I have drawn upon research conducted over the past 25 years at the archives and libraries thanked in the acknowledgements in my six previous works published by the MIT Press. Thanks to flexible and understanding editors, first Sara Meyrowitz and then Margy Avery, I was allowed to transform the focus of this book considerably as it went forward. And thanks to a fortunate combination of local circumstances, for the first time in twenty years, in 2007 I had a research assistant, Thomas Johansen. Together, we wrote a short book on a different topic, and for this book he did library research, created order out of the documents I had assembled, and took on some routine tasks that gave me more time to write.

Several institutions provided me with the opportunity to speak about this research as it moved forward, including the Twenty-Fourth National Regulatory Conference (Williamsburg, May 11, 2006), Oxford University's Said Business School (May 26, 2006), Oslo University (September 29, 2006), the University of Munich (November 2006 and May 2007), SDU-Kolding (April 2008), the European Association for American Studies (Oslo, May 2008), the Department of the History of Science and Technology, the Royal Institute of Technology (Stockholm, May 2008), and the Society for the History of Technology (Lisbon, October 2008). A small portion of this work appeared in my paper "Are blackouts landscapes?" (*American Studies in Scandinavia* 39, 2007, no. 2: 72–84), and I thank the editor for allowing me to redeploy these materials here.

Helle Bertramsen Nye put up with my chatter about this topic and offered much sound advice. Many others drew attention to useful materials and offered encouragement, including Klaus Benesch, Miles Orvell, Richard Hirsh, Jeffrey Meikle, Alex Roland, Mark Luccarelli, Per Winter, and Leo Marx. The American Studies Writing Circle at SDU discussed a draft of chapter 6 with me and offered useful feedback.

I dedicate this book to the memory of my mother, Fern D. Nye, who from childhood on was my most generous reader. Her first years were spent in the pre-electric world of rural Pennsylvania, and she experienced her share of blackouts. At the end of her days, in 2008, she was casually using electric kitchen appliances, DVDs, email, a mobile phone, and many other gadgets that appeared during her lifetime. She knew firsthand what electrification had meant.

When the Lights Went Out

Introduction

Where were you when the lights went out? People ask this question because electrical blackouts are carved out of the normal flow of time. Anyone with electrical service has experienced at least one blackout, and major disruptions are etched in memory. But most people's memories of lesser blackouts fade as soon as the lights come back on. Power failures tend to be relegated to technical analysis, and seldom have been studied as social or cultural history.

I remember precisely where I was sitting in the Robert Frost Library at Amherst College when the Great Northeastern Blackout of 1965 began. I cannot pinpoint much else from that year with equal precision. I initially assumed that the disruption was temporary and local. I suspected it was a misguided fraternity prank, and I was a bit peeved because I was preparing for a midterm exam. That was more than 40 years ago, yet I can visualize the spot where I sat in the sudden darkness. Emergency lights guided me down the stairwell. Fortunately, the evening meal was already cooked in the college's cafeteria. My inconvenience, as I would read in newspapers for the rest of that week, was slight in comparison to that of millions of people in cities, some stuck in stalled elevators or subway cars.

Four decades later, blackouts became a subject for my research in other libraries, after an academic journey that can be summarized by the subjects of my books: the automotive revolution and the assembly line, Thomas Edison, General Electric, the history of electrification, the technological sublime, the history of energy consumption, and narratives of technological transformation. That research contributed to this volume, preparing me to understand the blackout in multiple ways: as a disruption of social experience, as a military tactic, as a crisis in the networked city, as the failure of an engineering system, as the outcome of inconsistent political and economic decisions, as a sudden encounter with sublimity, and as memory, aestheticized in photographs. One must understand the blackout in all these ways and more. The subject here is not simply power outages, but different social constructions of artificial darkness, whether due to warfare, strikes, accidents, shortages, market manipulation, terrorism, or the voluntary actions of environmental organizations.

The chapters of this book are roughly chronological. Chapter 1 explains how electricity became a part of everyday life and how the electrical system was developed into a grid that, by the 1930s, made large power failures possible. Chapter 2 looks at military blackouts before and during World War II, which had the effect of normalizing what was hidden: the brightly lighted urban landscape, which would return, as Vera Lynn sang, "when the lights come on again all over the world." Chapter 3 traces the postwar increase in electrical consumption and the expansion of the grid to supply it, which created a dependence that became strikingly evident in 1965. The public response to the Great Northeastern Blackout was as unexpected as the power loss itself, revealing a spontaneous capacity for kindness and civic solidarity that forms one of the book's sub-themes. Chapter 4 emphasizes that such responses cannot necessarily be expected, but that they emerge from and express particular historical circumstances. (In 1977, another blackout in a quite different context spawned a wave of arson and looting in New York.) Chapter 5 examines another form of power loss: the rolling blackouts that occurred with increasing frequency in the 1980s and afterward, many of them due to shortages of generation or transmission capacity but others (around the year 2000) due to energy traders' efforts to "game the system." Chapter 6 turns to blackouts caused by terrorist attacks, still only a latent possibility in the United States but common enough in other parts of the world (notably Iraq). Imagining and containing sabotage has become a full-time job for the Department of Homeland Security and for many employees of utility companies. It has also become a staple of novels and films. Chapter 7 examines "greenouts"—voluntary blackouts organized since 2007 by environmental activists concerned about resource depletion, species extinction, pollution, and global warming.

As the progression of topics suggests, the subject of blackouts intersects with various cultural concerns, including economic growth, war, energy supplies, labor unrest, social justice, consumerism, resource shortages, capitalism, terrorism, global warming, environmental degradation, and the quality of urban life. The arrested moment of each blackout provides a snapshot of the electrical system and of social relationships. Blackouts are breaks in the flow of social time, and examining a series of them reveals much about the trajectory of history. During most of the twentieth century, Americans doubled their domestic electrical consumption every 10 years, becoming so dependent on electricity that today life comes to a halt if the power fails. In these moments of stasis, Americans are forcibly reminded that they are not isolated individuals, but a community that now needs electrical wires and signals to bind itself together.

Much of what is written about blackouts assigns blame. Newspapers often blame utilities. Utilities blame government red tape and over-regulation. Regulators blame lack of inspection and lack of enforcement powers. Elected officials may blame utilities, another political party, or another country. Official reports, whether prepared by utilities or by government commissions, focus not on those who are unharmed although inconvenienced, but on the concerns of power station managers, public officials, and those who suffered monetary losses. Such publications give the chronology of each event, explain technical failures, blame specific actors, and assess economic costs. After the 2003 blackout that affected 50 million people in Canada and in the northeastern United States, a number of such reports appeared. A Canadian-American task force devoted 27 pages to "how and why the blackout began in Ohio" and 30 pages to how failures there cascaded from one system to another.[1] But that report contains almost nothing about the public's immediate response to the blackout, or about its subsequent behavior. Though it is useful to know whether particular power failures were due to lightning, to human error, to poor maintenance, to squirrels chewing on power lines, or to inadequate generating capacity, the social and cultural history of blackouts receives little attention in such accounts. They treat each blackout as a unique event that is destined to become a legal case governed by the laws of contracts and torts.

The reports written after major power failures establish many facts but provide little insight into the social meaning or the historical significance of blackouts.[2] They seldom examine the often intense and unforgettable experiences of ordinary people, and they do not consider their non-technical interpretations of what a power failure meant. For those caught in a blackout, technical assessments do not capture the experience. In 1942, Harold Ross of *The New Yorker* described riding on top of a Fifth Avenue bus when a wartime blackout began. On Seventy-Fourth Street the bus had to pull over and shut off its headlights. "The lights in the buildings in the Park, and on the street corners went out and about 30 seconds later the traffic lights did too. Passengers were told to put out their cigarettes." One more streetlight was turned off, and this "left only the red beacon on top of the RCA building, and the moon. The silence was the big surprise of the blackout, the darkness discounted."[3] From many such accounts, one can piece together the shared characteristics of all blackouts and also see what behavior is unusual or atypical.

In exploring the empirical material, I use several approaches from the humanities and the social sciences. Because blackouts are not only technical malfunctions but also social events, economic problems, and political emergencies, an interdisciplinary approach is unavoidable. I focus only briefly on mathematical models and computer simulations of blackouts. I do not try to summarize the economic literature on utilities. I have almost nothing to say about utilities' metering and pricing policies, for example. Nor do I attempt to compress a course on electrical engineering into the book. But I endeavor to understand enough of these subjects to avoid errors when touching upon them. To explain the development of the electrical generation and transmission system, I have relied on Thomas Hughes's theory of how systems are built and how they achieve technological momentum, and on Richard Hirsh's analysis of how American electric utilities were deregulated and restructured.[4] To understand why blackout failures occur, I turned both to Charles Perrow's concept of "normal accidents" (which become less foreseeable as a technical system is more tightly interlinked) and to the research of specialists in control theory, including John Doyle, Dusko Nidcic, Ian Dobson, Daniel Kirshen, Benjamin Carreras, and Vickie Lynch.[5] However, none of these approaches has much to say about the social experience of a blackout. Michel Foucault's concept of heterotopia suggests that the blacked-out city becomes a new kind of social space.[6] Victor Turner offers a way to understand the blackout as liminal social time.[7] Lee Clarke's work on "fantasy documents" illuminates organizational preparation for crises such as blackouts. J. B. Jackson's studies of landscape suggest ways to think about the long-term implications of high-energy spaces.[8] This is by no means a complete list of the scholars who inspired the present work, but these are the most important. Each theory is briefly introduced as the story moves between technical, social, political, and cultural history. Specialists often want theory to be highly visible, like scaffolding covering a building. I prefer to build it unobtrusively into the argument (with documentation in the notes), so that the book will be accessible to more readers. After all, just about everyone with electrical service has experienced at least one blackout and is likely to experience another. My subject in this book is this historically new collective experience.

The Lindbergh Beacon at night, Los Angeles, February 15, 1929. General Electric Corporation.

1 | Grid

Few people in the United States or in Europe remember life before electric light and power, unless they are over 80 years old or grew up in a remote area. Most know only an artificial darkness that is fogged with electric light. Even in the late nineteenth century, people living away from the more polluted cities, such as London or Pittsburgh, had a good view of the night sky. Particularly on moonless nights, they knew profound darkness, and they could see thousands of stars. For several generations, however, the heavens seen from American and European cities and suburbs have been opaque. The air is filled with fine particles and pollutants, and electric lights produce "sky glow" that washes out the fainter stars and dims the Milky Way. Light pollution and air pollution make it impossible to see more than the brightest stars and planets with the naked eye. Even in a comparatively remote location such as the Grand Canyon, where the heavens seem spectacularly visible, "sky glow" and air pollution from Las Vegas partially obscure even the clearest night.[1] For most people, the night sky has become a smudged and meaningless background. As Henry Beston observed in 1928, people no longer feel comfortable with darkness:

[We] now have a dislike of night itself. With lights and ever more lights, we drive the holiness and beauty of the night back to the forests and the sea; the little villages, the crossroads even, will have none of it. Are modern folk, perhaps, afraid of night? Do they fear that vast serenity, the mystery of infinite space, the austerity of stars? Having made themselves at home in a civilization obsessed with power, which explains its whole world in terms of energy, do they fear at night for their dull acquiescence? . . . Today's civilization is full of people who have not the slightest notion of the character or the poetry of night, who have never even seen night.[2]

This change in human experience began with gas lighting in the early nineteenth century and became more significant with the introduction of commercial electric lighting in the 1870s.[3] Landscapes are cultural constructions that emerge from political, economic, and aesthetic decisions, as is particularly obvious in the case of the electrical landscapes that emerged after 1890—notably New York's Broadway (popularly dubbed the "Great White Way" as early as the 1880s) or, in recent times, the Las Vegas casino strip.[4] Night views of a city seen from the top of a skyscraper began to seem normal by 1920. (See figure 1.1.)

Electric landscapes often developed in imitation of world's fairs and amusement parks.[5] As advertising, floodlighting, and street lighting spread through urban society, salesmen were able to ratchet up sales of electrical signs to businessmen eager to maximize the visibility of their enterprises.[6] An unintended result of their competition was a brilliant landscape of light that by the 1920s had transformed urban space into a new form of the sublime that hid the night sky. Such spectacles were gradually accepted as "natural." At first the electrical sublime did not seem to carry within it the possibility of breakdown or radical dysfunction. Rather, electrical displays erased the heavens and transformed the night into a canvas to be painted with lights,

Figure 1.1
Night view of lower New York City from the Metropolitan Tower,
c. 1920. Library of Congress.

bringing out monuments and impressive buildings and surrounding them with spangles of man-made stars.

In Immanuel Kant's classical formulation, the sublime object
makes a powerful first impression and is so extraordinary that
with each subsequent encounter it seems equally stirring and
absolutely great.[7] Niagara Falls or the Grand Canyon does not
lose its power with repeated visits.[8] Similarly, the electrified city
seems to lure observers repeatedly to ascend skyscrapers or to
walk into Times Square or Piccadilly Circus to experience the
dazzling tapestries of flashing lights.[9] This illuminated cityscape
was not planned; it arose despite efforts by the City Beautiful
Movement to regulate the colors and the intensity of lighting
(with little success except on Manhattan's Fifth Avenue, at some
exposition sites, and in a few other locations).[10] The kinds of
signs that were banned from Fifth Avenue predominated in the
commercial centers of most American cities. New York's Times
Square and its lighted skyscrapers became emblems of the urban
night, often reprised in films such as *Manhattan* and *Sleepless
in Seattle*. The night cityscape, whether that of Pittsburgh, San
Francisco, Boston, Las Vegas, Chicago, or Washington, is endlessly reproduced on postcards. Los Angeles is often represented
as a glittering sea of lights. The night skyline has become the
signature image of the metropolis, a defining landscape of
modernity.

Human beings are no longer awed by the immense changes
electrification made possible in lighting, manufacturing, transportation, and domestic life, and few are concerned that nations
have become so dependent upon electricity—perhaps because
it now seems to have been always already there. The wall switch
and the light socket seem natural, a power failure unnatural.
People notice electricity only in its absence. When camping in

the wild, they may briefly contemplate a darker world that once was normal. "Before electricity," George F. Will once noted in *Newsweek*, "conditions of life were more like those of Julius Caesar's day."[11] Had Caesar returned from the dead and visited George Washington, he would have understood Washington's means of transport and his sources of light and power, even after 1,800 years.[12] But during the nineteenth century the steam engine, the electric generator, and the gasoline engine brought decisive changes. Caesar would not have understood how trains or the first automobiles could move without apparent external aid. He would have wondered at elevators racing up inside skyscrapers, at cities bursting with brilliant illumination, at the productivity of vast factories, and at steel-hulled ships, machine guns, and other new armaments. Such innovations also amazed nineteenth-century Americans, who celebrated new railroads, skyscrapers, powerhouses, dams, and illuminated city skylines—often rather uncritically—as exemplars of the technological sublime.[13]

These technical "wonders" had unintended consequences: huge demands on natural resources, pollution of air and water, envelopment in an increasingly artificial environment, and, occasionally, catastrophic system failures. Blackouts were among the unintended consequences. Between 1880 and 1940, the electrical system invented by the team at Thomas Edison's laboratory, and refined and improved by many others, spread across America.[14] Scarcely anyone considered the possibility of comprehensive power failures. (A blackout presupposes an electric grid that did not yet exist.) Rather, the public celebrated the brilliant displays of illuminated cityscapes.[15] It takes an effort to recapture the sensibility of people living at the beginning of electric lighting. Until the late 1870s, almost all light had been some form of fire, and the night had always been dark. Intensive lighting is a historical anomaly, and it remains so for many living in poor countries. In *At Day's Close: A History of Nighttime*, A. Roger Ekirch recovers the rhythm of daily life in the early modern era, when darkness was still normal.[16] In the pre-modern night, society did not close down at dusk, but people divided activities into those that could be performed only during the day and those that were consigned to the dark. They had a firm sense of two different realms. As societies adopted electrification, however, the abolition of darkness began to seem normal. People began to expect temporal homogeneity so that all activities were possible around the clock.

For most of history, darkness and light alternated in a rhythm that varied with the seasons, and that imposed a structure and limits on existence. Breaking the cycle suggested breaking the bonds of nature. The first world's fairs closed at dusk, but after 1881 a fair or an exposition could remain open after dark. Lavish lighting at the expositions in Chicago (1893), Buffalo (1901), St. Louis (1904), and San Francisco (1915) demonstrated the dazzling possibilities.[17] With power and light perpetually available, factories could run continuously. Major thorough-fares, such as Atlanta's Peachtree Street (figure 1.2), were lined with ornamental lights, and stores were kept open at night. Outdoor sports, once confined to daylight hours, could be played "under the lights." Popular magazines described a future that would be constantly illuminated.[18] Edison declared that electrification, as it sped up society, might even hasten human evolution.[19]

Such predictions contrasted sharply with Victorian realities. In 1881, when utility companies began to install systems for generating and transmitting electricity, they did not immedi-

Figure 1.2
Ornamental lighting, Peachtree Street, Atlanta, June 2, 1920. General Electric Corporation.

ately displace familiar sources of light, power, and heat (notably wood, gas, kerosene lamps, and candles). Even a generation later, some architects advised clients to install both gas pipes and electrical outlets. Only 5 percent of houses had installed wiring in 1905,[20] when electricity was primarily for public display and commercial convenience. Electricity was not yet deeply embedded in everyday life. Although early electrical systems often failed, there were no blackouts. Major buildings, such as department stores, large hotels, and the New York Stock Exchange, often had stand-alone generating plants. Every home or business had multiple sources of light and energy. Between 1880 and 1905, this patchwork of local services might fail individually but not collectively.

During its first quarter-century, the electrical system moved from a speculative investment to an established innovation. In *Networks of Power*, Thomas Hughes notes that after invention and early installation (1875–1882) the new system was transferred to other regions.[21] As the system grew and attained technological momentum[22] after 1895, uniform standards were adopted. A large bureaucracy, thousands of workers, and millions of consumers became committed to the same alternating-current (AC) system, with 110-volt lines, light bulbs that screwed into sockets rather than sliding into them, and standardized frequencies.[23] The mature system consisted of much more than power plants and transmission lines; it also included the wiring of millions of buildings, appliances and their manufacturers, local electricians, educational institutions, research laboratories, and regulatory agencies, forming what Hughes calls a "socio-technical system." It had powerful momentum because of its "institutionally structured nature, heavy capital investments, supportive legislation, and the commitment of know-how and experience."[24]

As this socio-technical system expanded, utilities created interconnected local and regional grids. Before then, a power station that suffered an accident or one that was struck by lightning might stop producing electricity for a long period. For example, the utility company in Hartford lost power on June 5, 1890, and did not restore full service until late autumn. From such experiences, engineers quickly discerned the advantages of linking systems into regional networks.[25] For example, northern New Jersey had many small generating plants in 1906. Each was an independent unit, and none could produce more than 1,800 kilowatt-hours (kWh). *Electrical World* explained how the new Public Service Corporation (a private company despite its name) might consolidate these units: "[T]he stations are located in the wrong places, the several (distribution) networks are not so disposed as to help each other, the railway lines do not transfer at the right places, if at all, and the fares are both high and ill-adjusted." A consolidated company could "step in, put the heterogeneous plants into touch with each other, replace the inefficient ones, lay out the feeder system for the whole territory, unify the transportation systems so as to get efficient transfers and through lines, and in general to give better service while itself gaining from lower operating costs and smaller general expense." Welding the distribution network together to form a local grid would give "security against breakdown, both of lights, and traction lines."[26] Once established, Hughes argued, such an electrical grid was "less shaped by and more the shaper of its environment."[27]

After the first generation of inventors and entrepreneurs decided where to build the grid and selected its specifications and components, it became difficult for later generations to choose alternative arrangements.[28] What began as a voluntary choice became a requirement. As electrification gained momentum, utilities that did not upgrade their systems voluntarily were forced to do so by consumer demand. In 1919 and 1920, Philadelphia's antiquated grid could not handle that demand, and frequent breakdowns caused blackouts in the city's center. Theaters briefly had to resort to oil lamps. The chief engineer put the blame on an outmoded Edison DC plant with inadequate connections to the rest of the system. Not certain that it should abandon direct current, the utility company dithered for months before opting for an AC system.[29] After central Philadelphia was linked to the surrounding grid, power outages declined sharply.

As utility companies consolidated and expanded, they increased their service to homeowners. Whereas in 1905 only 5 percent of all urban houses and apartments had electricity, by 1930 more than 90 percent did.[30] Initially electricity was used chiefly for light, but soon people began to acquire appliances. Many housewives first bought an electric iron, which was cleaner and easier to use than one heated on the stove, and far more agreeable during warm weather. Soon they adopted electric vacuum cleaners, clothes washers, toasters, and waffle irons, and to a lesser degree electric stoves and mixers. Utility companies portrayed the wired house as a "home of a hundred comforts" (figure 1.3). Electric refrigerators were still expensive and unreliable, however, and iceboxes remained the norm.

Figure 1.3
Illustration for booklet "Home of a Hundred Comforts" (March 1925). General Electric Corporation.

Nevertheless, by 1930 urban Americans had woven electricity into many habits and structures of daily life, and it had insinuated itself into their aspirations. People expected frequent product improvements and entirely new appliances, and young urbanites could not imagine a home without electricity.[31] In all, it had taken 50 years from Edison's first demonstration of his electrical lighting system in 1879 until its domestication.[32]

Few realized that building the electrical distribution system created not one but two new social realities. The predominant one was the interconnected world of the grid, which encouraged linkages between technical systems, creating a network of networks. But as the grid became the backbone of everyday life, a second, latent social reality also emerged: the blackout. As Stephen Graham and Nigel Thrift have noted, breakdowns "are not aberrant but are a part of the thing itself. To invent the train is to invent the train crash, to invent the plane is to invent the plane crash."[33] The grid was inseparable from the blackout. Two contradictory social realities were latent within the electrical system, depending on whether or not it worked. Though enormous benefits flowed from it, by 2004 the grid's many failures cost the United States as much as $79 billion a year.[34]

It is hard to anticipate how an invention will be used. Consider Edison's phonograph. In 1878, it was a technical wonder that amazed paying crowds. Edison expected it to be used to dictate business correspondence and to preserve voices of loved ones for posterity.[35] It took more than a decade to discover that the public would use the phonograph primarily to listen to music.[36] What would the public do with Edison's electrical system? In 1879 his laboratory team created not only an incandescent light bulb but also the devices required to make it useful, including wall switches, sockets, insulated wiring, underground cables, and generators. No one could entirely foresee the vast potential of such an energy system, but within a generation thousands of new inventions would be plugged into it. People used electricity to heat baby bottles, to light Christmas trees, to toast bread, to regulate room temperatures, to operate traffic signals, to drive trolley cars, to light streets, to fan the air, to run assembly lines, to pop corn, to raise and lower elevators, to maneuver cranes, to make coffee, to light cigars, to air condition buildings, to project motion pictures, and indeed to do anything that required motive force, heat, cooling, light, or the transmission of messages.

As inventors created more ingenious devices, consumers' demand for electricity doubled roughly every ten years from 1900 until 1970. Yet combinations of machines, invented separately, can create unstable systems when linked together. As the demand for electrical power grew, and as utility companies created a vast electrical grid, an unanticipated problem emerged: blackouts. At first, power outages were quite local, usually brief, and far more frequent than present-day consumers would accept. There were many small power providers. Factories, streetcar lines, and other enterprises owned stand-alone generating systems and sold power and light to other nearby businesses. A failure on one side of a street often had no effect on the other side. The streetlights might go out, but the streetcars (which used a separate DC power system) still ran. Furthermore, during the early decades of electrification most businesses and homeowners had alternative light and power ready to hand in case the electricity failed.

Because no local utility company could supply the electricity that the Chicago Columbian Exposition of 1893 required, a separate plant was built at the fairgrounds. The exposition's spectacular lighting, electric fountains, electric railroad circling the grounds, and enormous Electricity Building all pointed toward an electrified future. The Westinghouse exhibit demonstrated the new technology of alternating current, which could be transmitted economically over long distances. The further direct current was sent, the larger and more expensive the copper wires had to be. Utilities therefore built DC generating stations in the center of a ring of customers, as was the case with the first Edison plant, the Pearl Street Station in lower Manhattan. DC plants had few interconnections, because transmission over long distances was uneconomical, and therefore no large-scale grid was practical.

Thomas Edison's former private secretary, Samuel Insull, led the transition to large AC systems. Insull also championed the idea of natural monopoly. AC and natural monopoly were the technical and political sides of the same coin. Yet when Insull left Edison's employ to head a Chicago utility company, few imagined comprehensive urban interconnections, much less a regional grid. In 1898, the city of Chicago had eighteen electrical utility companies, each of which competed with gas companies in its local market. Rural areas of the United States did not have electrical connections at all. But as the system grew, lines were linked. In most cities and towns, as in Chicago, one company gradually acquired the others. Consolidated utilities achieved economies of scale. It was efficient to produce electricity with large steam turbines or hydroelectric dams. Insull persuaded Chicago factories and traction companies to give up their small power plants and rely on his utility. He installed some of the biggest generators in the world. By 1912 their average output was 37 megawatts, seven times the London average. Because of such economies of scale, Chicagoans paid less than half as much as Londoners for each kilowatt-hour they used.[37]

Early powerhouses had steam engines. Insull replaced them with more efficient steam turbines, and he pushed General Electric to create larger and more efficient units. He knew that scaling up generation was crucial to making his monopoly grow. Larger turbines lowered the cost of producing a kilowatt-hour, and Insull could then reduce rates to attract more homeowners. The larger Insull's system became, the more efficient it was, and the more it justified running utilities as monopolies. Year after year, Insull saw other utility managers at meetings of the National Electric Light Association and showed them how economies of scale could bring down the rates paid by consumers and yet increase profits. In the rate structure that Insull and his colleagues developed, homeowners paid a premium, while large customers paid low bulk rates in exchange for guaranteeing the base load.

After about 1910, it appeared to Americans that electrical utilities, by their nature, worked best when run as local monopolies. Rather than let two or more companies build electrical lines all over town, surely it was less expensive to have a single system. Rather than build many small generating plants, surely it was better to have a few large ones. Duplication was wasteful. The utilities promoted themselves as "natural monopolies" responsible for all aspects of a region's power generation and delivery. The courts recognized "natural monopoly" as a legal doctrine.[38] They held that an electrical network, by nature, had to be comprehensive and integrated, uniting in just one company the responsibility for production, distribution, and sales. State governments also embraced "natural monopoly" and its logical corollary, regulation. To ensure that prices were fair and services comprehensive, every state created utility commissions, or oversight boards. To guard against corruption, shoddy engineering work, and price gouging, these commissions monitored utilities and set prices. With their oversight, the American power system doubled and redoubled as alternating current extended the system's range and as ever-larger turbines and dams reduced production costs. Residential customers paid about as much for a year's electricity in 1937 ($24.81) as they had in 1913 ($22.97), even though they consumed three times as much (793 kWh vs. 264).[39] Consumers, shown new uses of electricity in demonstration homes (figure 1.4) and at world's fairs, rapidly increased their use of electricity use.

Figure 1.4
"New American Homes," such as this one in Atlanta (July 1936), displayed the latest electrical appliances to an eager public. General Electric Corporation.

One reason costs fell was that the grid's interconnections enabled utilities to sell surplus electricity. Because they could buy from one another, utilities did not have to keep as many reserve power plants on standby to meet high demand. As in Insull's Chicago system, regional grids began with local interconnections. After 1900, a large system emerged in central Canada and upstate New York, linked to the power plants at Niagara Falls. During World War I, unprecedented demands for electricity further demonstrated the need for regional systems. By 1920, legislative proposals for grids emerged both at the state level and at the national level. Governor Gifford Pinchot of Pennsylvania wanted his state to take over power generation and build large plants close to coal mines. Pinchot planned to reduce utilities to middlemen that transmitted power to the consumer.[40] Such proposals were often branded as socialistic, however, and whatever their technical merits they did not succeed politically.[41] Instead, private corporations collaborated to build regional grids. For example, New Jersey and Pennsylvania were tied together in a regional grid that included a huge new hydroelectric dam across the Susquehanna River.[42] System building took place in all parts of the United States during the 1920s. In 1923, Harvey Couch, a young Arkansas entrepreneur who had built up a telephone company and sold it to the Bell System, purchased four urban electrical utilities in Mississippi—in Jackson, Vicksburg, Columbus, and Greenville—and tied them into a network. His Mississippi Power and Light Company achieved greater efficiency through integration, and between 1924 and 1930 its electricity sales rose 500 percent.[43] In Denver, the entrepreneur Henry Doherty won customers by undercutting competitors' rates. Like Samuel Insull, Doherty understood economies of scale. He hired a large sales force and pushed expansion virtually from the day he arrived in Denver in 1900. Success generated capital for further expansion, and by 1935 Doherty controlled 183 interlinked electrical companies in eighteen states.[44] Likewise, California utilities expanded and merged into a comprehensive state grid during the 1920s.[45]

Until the 1930s, the United States was moving rapidly toward an all-encompassing grid largely controlled by private monopolies. Much of it was owned by holding companies in complex business arrangements that crossed state lines. Had the process gone unchecked, a few monopolies probably would have controlled a national system. But President Roosevelt wanted more competition in the electrical industry, and his administration built federal dams that established an alternative to private power. The Tennessee Valley Authority (TVA) and other federal power suppliers provided independent data on the actual costs of generation and transmission. This information was sorely needed by the new Federal Power Commission. Established by the Federal Power Act of 1935, that commission broke up some interstate power monopolies and weakened the grip of holding companies.[46] At the same time, another federal program, the Rural Electrification Administration (REA), extended service to the not-yet-electrified countryside.[47]

There was no federal department of energy, however, until the late 1970s. Until then, the federal government made *ad hoc* energy plans to solve particular problems. Yet it did assist the development of grids, as its agencies and departments built new high-tension lines that connected regions. Notably, there was a need for power lines to link Los Angeles and Phoenix to Hoover Dam, to link Portland and Seattle to the Bonneville Dam, and to link Knoxville and Chattanooga to the many TVA dams. Indeed, demand for electricity grew during the

Great Depression, and construction on the Grand Coulee Dam (figure 1.5) and other facilities continued around the clock.

In the late 1930s, construction of links in the grid temporarily came to a halt as a result of internal bickering in the Roosevelt administration and opposition from private utilities, which feared erosion of their autonomy.[48] However, World War II created a surge in energy demand that made new and better transmission lines a national priority. Enormous amounts of electricity were used to make bombs and fertilizer, to make steel and aluminum for battleships and airplanes, and to run factories around the clock. During World War II, the REA continued to

Figure 1.5
The Grand Coulee Dam under construction in 1941. Library of Congress.

spread into the countryside. In the Western states, the Bureau of Reclamation built dams and irrigation systems and extended and improved transmission lines. The Army Corps of Engineers built many dams and constructed transmission lines to the Department of Defense's far-flung facilities. Though these simultaneous efforts, the lattice of interconnections continued to spread. Although the federal government would not co-ordinate national energy policy until the last decades of the twentieth century, its many activities nevertheless accelerated the integration of the electrical grid. Furthermore, growth was self-reinforcing. As electricity prices fell, more applica-tions became economically attractive, increasing demand and making possible larger and more efficient power stations. These in turn lowered the price still further, driving growth ever higher.

As the electrical system extended into every part of daily life, it became the network that underlay all other networks. From water systems, railroads, gas lines, sewer systems, and telephone exchanges to the stock market ticker, electricity became essen-tial to control systems that included switches, pumps, valves, sensors, alarms, warning lights, transmitters, switchboards, fuses, and monitors. As electricity wove networks together, power failures became less and less tolerable, because they shut down the entire infrastructure. In the late nineteenth century, small power outages had been frequent but had affected only small parts of the piecemeal electrical system. As reliability and integration increased, so did vulnerability.

Service was continuous, but both utilities and regulators assumed that short power outages due to natural causes were unavoidable. After decades of dealing with ice storms, lightning, and hurricanes, utility linemen and maintenance crews knew how to make rapid repairs. The rule of thumb was that "loss of load on one day in 10 years, or 2.4 hours in the average year" was acceptable.[49] This did not mean that all customers lost service for 2.4 hours each year, but that over 10 years it was deemed acceptable for a system to fail any one customer for a total of 24 hours. These calculations had a name: "value of lost load" (VOLL). The more robust the transmission lines and the greater the reserve generating capacity, the smaller the VOLL was likely to be.[50]

Hydroelectric power is an especially attractive reserve, as the cost of generation is much the same regardless of when one uses it, and it can be turned on and off more quickly than a large coal-fired generator. Indeed, the low cost of hydroelectric power was an incentive to building many of the major connections in the grid, notably the lines radiating from Niagara Falls, Hoover Dam, the Grand Coulee Dam, and the TVA. Yet in 1965 reliance on such long high-tension lines led to a major blackout of the entire area between Toronto and New York City. Such a blackout was unimaginable before it occurred. It was an unintended consequence of building the grid during the previous 75 years, after alternating current had made long-distance transmission feasible.

Since 1965 there have been other major blackouts, despite continual efforts to prevent them. The vast electrical network— the largest machine in North America—-is occasionally unstable, with failures every year that cut power to more than 10 million people. "Like that of many Third World countries," one Californian complained in the *Wall Street Journal* in 2001, "California's electrical grid can now fail with little notice."[51] A blackout in India affected 220 million people in the same year, showing that a gigantic failure is possible.[52]

The Electric Power Research Institute has estimated that blackouts cost American businesses as much as \$100 billion per year.[53] Electricity has become the economy's lifeblood. A national collapse in the United States would be exceedingly unlikely, because the grid is divided into regions with firewalls between them, and because electricity is produced at thousands of different sites. Portions of the system can and do fail, often from natural causes. (Squirrels cause more than 1,000 local blackouts a year by gnawing on cables.[54]) No amount of planning or expertise can entirely prevent outages, though it helps if there is excess generating capacity and redundancy in the grid. Yet there are many human reasons for blackouts, including excessive demand for power on hot summer days, improperly installed or poorly maintained equipment, failure to cut back trees beneath power lines (which sag as they heat up), fuel shortages, errors by operators of power plants, glitches in monitoring systems, and the refusal of citizens to allow construction of new plants and transmission lines.

Some utilities, some states, and some countries have more reliable and more abundant power supplies than others. In the middle of the twentieth century, the United States had the world's most comprehensive and most reliable power system. No more. Today Europe has fewer and shorter blackouts, in part because its system has greater reserve capacity. Nevertheless, blackouts still occur in Europe. Notably, on September 28, 2003, 57 million Italians suddenly lost power in the middle of the night.[55] Why do some countries have more reliable power than others? For a start, countries' power systems are by no means identical. When current comes out of a socket, the consumer cannot tell if it is made by wind turbines, solar panels, falling water, an atomic reactor, or the burning of oil, coal, gas, peat, or waste. Even if streams of electrons are identical, the systems that produce them are not. Over 98 percent of Norway's electricity comes from falling water. Sweden relies on nuclear power plants for almost half of its electricity. Nearly 90 percent of France's electricity is from nuclear plants. More than 75 percent of Australia's is from coal-burning plants. Most countries rely on multiple forms of generation, typically including hydroelectric dams built in the early decades of system building, coal-burning power stations built later, nuclear plants typically constructed after 1970, gas-fired plants built more recently, and, newest of all, renewable forms of generation (often windmills, less frequently solar panels or geothermal wells). Each composite system is tied together to form a grid that must constantly balance supply and demand. Every regional network is linked to larger grids.

Some forms of generation are more ephemeral than others. Coal or gas can be reliably stockpiled for burning. Rainwater can be impounded behind a dam. But wind and sunshine must either be made into electricity immediately or transformed into some other form of energy, such as compressed air stored underground or water pumped to a higher elevation. Unlike most industrial products, electricity must be used the second it is produced. Consumer demand fluctuates in a daily rhythm that peaks once in the morning, remains high during the day, peaks again in the late afternoon or early evening, and tails off to its lowest levels after midnight. Utilities must make educated guesses about how much power consumers will need, and must continually adjust their production and their purchasing of power. They do not use all their generating sources constantly. To meet the minimum constant demand, they choose a form of power that can be reliably produced at low cost, reserving more expensive forms for periods of peak demand.

Imagine how this might work if an individual house had to produce its own electricity. A simple solution (and indeed the one chosen by utility companies in their first decades) would be to install an automatically regulated coal-burning furnace to produce steam to drive a dynamo. This would require a coal bin in the basement, regular coal deliveries, and some unpleasant smoke. A cleaner alternative would be to burn natural gas. A householder who did not like depending on a single source could also install a windmill or some solar panels. However, each additional source of power would force the homeowner to spend more time running a more complex system. The owner would also discover that, rather than run a dishwasher, a clothes washer, a vacuum cleaner, an oven, a computer, and other appliances simultaneously, it is better to spread the demand out so that the system runs at close to a steady rate day and night. Another way to balance the load would be through neighborhood agreements to exchange power back and forth at an agreed-upon rate.

In practice, few homeowners want to run their own electrical systems, although some wealthy families did in the 1880s and the 1890s. Today, except for pleasure boats and remote vacation cabins, most Americans rely on the electrical grid. In 1960 it was the most advanced machine of its kind, and half of its components were less than 10 years old. But by 2005 the grid was a patchwork of old and new elements that badly needed an overhaul. Many generating plants had been constructed before 1970, and they did not have state-of-the-art equipment. In New York and Philadelphia, some underground cables laid in the late nineteenth century were still in use in 2000.[56] By 2002 the Department of Energy recognized that this aging infrastructure was "in urgent need of modernization," and in 2005 the American Society of Civil Engineers warned that "the existing system was built for local needs, and is struggling to meet the demands of a global system brought about by deregulation."[57] Furthermore, the transmission system had outmoded control rooms and insufficiently trained employees. During the blackout of August 14, 2003, which affected more than 50 million people from Ohio to the East Coast, some operators had trouble reading and interpreting the dials and gauges. They had not been adequately trained in simulations.

Even with new equipment and better-trained technicians, however, the power grid continues to break down, just as it did half a century ago. Consumers usually have a simple idea of why blackouts occur, often assuming that the distribution system is faulty. In practice, however, top-of-the-line technical equipment still can contribute to a blackout. Paradoxically, even if each part of the vast grid is upgraded, the system as a whole may become less reliable. As a technical system increases in complexity, apparently insignificant flaws and inconsistencies accumulate until an unforeseeable combination of circumstances can trigger a failure. After examining accidents of many different kinds, the organizational theorist and sociologist Charles Perrow concluded that they are inevitable in the tightly coupled and complex systems typical of electrified societies. In short, blackouts are "normal accidents" that can occur in many different and unforeseen ways.[58] Despite their normality, they are unpredictable, because it is not possible to know where a weak link, a faulty component, or a design flaw may be in so large a system.

Even when a blackout is apparently a random event, it is a cultural disruption. It is not just the electrical system that breaks down; the social construction of reality breaks down too. People must improvise temporary solutions to new problems, and often they must do this with strangers. Furthermore, the serious consequences of the disruption have increased with each passing decade, because electrical networks have become essential to more aspects of everyday life. In 2003 New Yorkers were dismayed to find that among the thousands of devices that did not work in a blackout were mobile phones, gasoline pumps, escalators, automated teller machines, and pumps in the water system. As the use of electricity increases, the response to a power failure changes from one blackout to the next, revealing how society is becoming more and more dependent on electric power and communications. A blackout unmasks the illusion that the city's appearance is "natural" or "normal" and confronts each individual with specific difficulties and with the larger challenge of making sense of the crisis as a whole.

Blackouts have many causes. In warfare, they are a form of intentional subterfuge. In labor conflicts, blackouts may be a strike weapon. In peacetime, blackouts are usually (though not always) accidents. Yet a squirrel or a terrorist can also cause one. The New York blackout of 2003 was a much different event, with a far different public response, than a blackout that occurred in 1936, in part because dependence on the electrical system had increased, but also in part because the imagined causes were more numerous. How have blackouts changed over time? How have people behaved when they occurred? To answer such questions one must leave utility hearings, corporate boardrooms and government offices to see what was going on in theaters, in airports, in skyscrapers, in stalled subways, and in the many environments that cannot function without light and power. As the professor of communication Susan Leigh Star notes, "the normally invisible quality of working infrastructure becomes visible when it breaks: the server is down, the bridge washes out, there is a power blackout."[59] Most immediately, every person caught in a blackout must redefine the potential uses of public space, and must improvise to some degree. This book is about what happens when the world's appearance suddenly darkens and time seems to stop. In 2005, the poet Lawrence Ferlinghetti argued that "what's called the dominant culture will fade away as soon as the electricity goes off." He identified the electrified world with the evasion of immediate reality: "In the 60s there was a famous slogan, 'Be Here Now.' . . . Today, with the telephones, the fax, the Internet, the whole schmear—the slogan you have today is 'Be Somewhere Else Now.'"[60] Ferlinghetti's critique has roots in European literature and philosophy, as well as in American Transcendentalism. Henry David Thoreau lived before most of the electrical inventions, but he cast doubt on whether new technologies are always an advantage and questioned whether the telegraph was really an improvement.[61] Likewise, in Max Frisch's 1957 novel *Homo Faber* an electrical engineer in Latin America installs electric power stations, which help to "eliminate the world as resistance." Electric light eliminates the night, air conditioning eliminates climate, and electric devices replace physical labor such as pumping water or grinding wheat into flour. The collective result of such technologies is that people lose their direct experience of the world. Similarly, Martin Heidegger and other philosophers have argued that in highly technological cultures people cease to feel a connection to the natural world. Human beings are part of nature, but they often use modern technologies to arrange their lives so that they scarcely experience it. One of Frisch's characters complains of "the technologist's mania for putting the Creation to a use, because he can't tolerate it as a partner, can't do anything with it; technology as the knack of eliminating the world as resistance, for example, of diluting it by speed, so that we don't have to experience it."[62] In this view, as the electrified domain mediates experience, it erases direct knowledge of and contact with the environment.

A comprehensive blackout reverses this situation, thrusting people into contact with one another and with the "resistance" of the world, which electrical devices usually holds at bay. The networked city shuts down. The elevator and the water system no longer overcome gravity. Streetlights no longer turn night into day. Telephones, radios, and televisions no longer defeat distance. If the blackout lasts long enough for smoke and pollution to blow away, the stars become fully visible again. The individual is cut off from virtual worlds and from distant relatives, friends, and colleagues. In a blackout, one suddenly must live in the here and now. With each passing decade, the disorientation caused by confronting an unelectrified world becomes more profound.

The *New Yorker* cover reproduced here as figure 2.1 suggests how thoroughly electricity had been woven into the city's self-perception by the autumn of 1930. It depicts Manhattan at night seen from an airplane, revealed as a vivid grid of lighted streets and skyscrapers. The biplane's black double wing on the left side establishes a pilot's point of view. Famous skyscrapers rise out of the electrical grid. Lights provide an illuminated map and a scintillating frame for the city's landmarks. Anyone familiar with New York can immediately locate the Brooklyn Bridge, the Brooklyn Navy Yard, and the slicing diagonal of Broadway. In 1930 few people had flown, but the aerial view was the modern, "air-minded" perspective, celebrating the technological mastery signified by the electrified metropolis and the conquest of the skies.[1] Within a few years, however, this celebratory vision would dim as the city's legibility from the air became a sign of vulnerability. In 1938, in a letter to the *Hartford Courant*, one resident of New York wrote: "The great size of our city with its numerous lights blazing all night make it an easy mark for bombing planes."[2]

The prospect of war in the air redefined the night. Darkness was no longer the backdrop for brilliant illumination; now it

Figure 2.1
Cover of *The New Yorker*, October 11, 1930. Used by permission.

was camouflage. Artificial darkness suddenly became desirable. This was a historic reversal. For 100 years, from the advent of gas lighting until the early 1930s, cities had increased their levels of illumination. They had done so because light freed commercial and recreational space for use at night, because well-lighted streets were easier to negotiate, and because illuminated cities felt safer and more modern. However, in the 1930s the electrical system and anything it illuminated became potential targets. The very word "blackout," used almost exclusively in show business during the 1910s and the 1920s, found a new meaning.[3] Before utilities developed regional grids, "blackout" was not used to describe a power failure in either New York's or London's *Times*.[4] Of course the electricity did fail occasionally, but between 1880 and 1930 outages remained local, at times confined to a city block or even one large building. Even after 1920, regional power outages remained unlikely, because technicians, not automatic relays, usually controlled the movement of power from one location to another. A power failure usually could not cascade from one utility to a neighboring grid. When the word "blackout" first was widely used, in the 1930s, it did not describe a widespread power failure, but rather the intentional act of hiding illumination. The word was also used to describe withholding information ("news blackout"). In each case, it meant the intentional control of appearances in order to prevent disclosure.

A military blackout disguised a fully functioning system. It demanded that the rationality and order of an electrified society continue under the cover of darkness. It imposed a mask. Elevators still worked, restaurants were still open, and traffic still moved efficiently. The military blackout was a comprehensive order imposed on the underlying electrical system, demonstrating two levels of mastery and efficiency. It also gave many citizens roles to play in drills that were staged as dramatic events. The goal was to render the city invisible, to erase the electrified landscape. This erasure could become a source of fascination during prewar blackout exercises that redefined urban space, hiding landmarks and disrupting familiar visual patterns. Citizens who habitually oriented themselves using an illuminated clock or building confronted a disorienting cityscape of shadows. Such defamiliarization, when expected as part of a prewar air raid drill, was less a hardship than a spectacle to be enjoyed.

Before aviation, war was largely confined to the front lines. But as early as 1911 an Italian reconnaissance pilot threw a few grenades down at Turkish troops in Libya. During World War I, planes carrying small numbers of bombs struck behind the front lines. A German plane bombed Dover in 1914, and zeppelins dropped bombs on London and Paris, but the deaths and the damage caused were less important than the psychological impact on civilian populations. The death toll from German air raids on Britain in all of World War I was only 557.[5] However, as the war came to an end the British were preparing a large-scale assault on Germany in which they intended to drop incendiary bombs on civilian populations.[6]

By the 1930s, rapid developments in aviation—notably the aircraft carrier—had made it obvious that any future war would include substantial bombing of cities, air bases, battleships, and supply lines hundreds of miles behind the front lines. The planes of 1917 had a limited range that permitted only short forays with a few bombs each. But two decades later an air force had far more lethal potential, flying further and faster with heavier payloads. All of France and England were within range of the east side of the Rhine, and planes flying out of Britain could hit targets anywhere in Germany. It became essential to confuse attacking enemy planes by disguise or misdirection. Even so, the first mention of a military "blackout" in the *New York Times* came as late as 1935: "Gibraltar will be plunged into darkness for an hour tomorrow night in an experimental blackout between 9 and 10 o'clock in connection with combined military and naval exercises."[7] Dousing the lights on stage had passed over to the theater of war. Gibraltar, guarding the entrance to the Mediterranean, was of obvious military importance. Practicing a "blackout" there seemed logical when Italian troops were invading Ethiopia and sectarian tensions were rising in Spain. Before 1935 was over, whole urban populations were forced to take part in blackouts on a much larger scale. Malta darkened its harbors and towns, and on the shortest day of 1935 Istanbul acted out an air attack, complete with planes dropping flaming fuses as substitutes for bombs and fire departments rushing to put them out.[8]

Blackouts swiftly moved from an element of military planning to a coordinated civilian exercise, not only in Western Europe but also in Japan and the Middle East. In 1937, a five-day blackout disrupted the lives of all of Tokyo's 5 million residents.[9] Theaters and other public venues closed, and families were required to cover their windows with thick paper. The Tokyo blackout mobilized 40,000 civilian volunteers in addition to the army and included practice ambulance runs, fire drills, and exacting tests of light control. To make certain that the artificial darkness was complete, in some districts the authorities turned off all power from the utilities, so as to be able to compare the intensity of the darkness before and after. In May 1938, all

Japanese cities were blacked out in anticipation of potential Soviet air attacks.[10] The British also organized for blackouts and installed sirens to warn of attacks in Alexandria and along the Suez Canal.[11] Aside from practice in logistics and in reorganizing urban life to function in the dark, these exercises taught citizens to regard themselves as targets, asked them to rethink their daily routines, forced them to ready their homes for possible bombardment, reconfigured the appearance of the city, and demanded that every man, woman, and child obey military orders.

In the United States, the small state of Connecticut alone organized 50,000 men and women as air raid wardens, 12,000 more as auxiliary firemen, and another 5,200 as auxiliary police.[12] In case of attack, they were to enforce the blackout rigorously. In New York City, 201,000 volunteers served as air raid wardens, not only patrolling the streets but also taking courses from more than 1,000 teachers in first aid and in how to deal with various situations that might arise. An additional 25,000 underwent 60 hours of training to be fire auxiliaries, 68,000 were ready to undertake demolition and decontamination work, and 10,000 doctors and 25,000 nurses were prepared to deal with medical emergencies. Half a million New Yorkers had clearly assigned roles to play in the event of an air attack.[13] At the same time, the military rapidly upgraded its anti-aircraft defenses, including powerful new searchlights at important bases (figure 2.2). Such measures had been largely unnecessary in World War I; however, as the size, power, and range of airplanes increased, much of North America became potentially vulnerable. The US Army Air Corps assigned Giulio Douhet's book *The Command of the Air* to its officers. Douhet argued that future wars would be won by bombing the enemy's factories, power plants, laboratories, and supply lines.[14] Douhet expected that precision bombing would

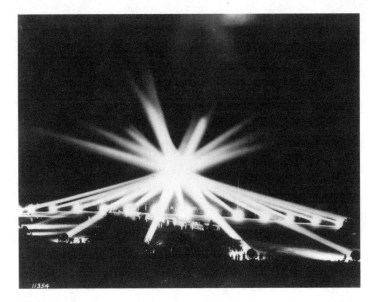

Figure 2.2
Sixty-inch searchlights displayed on the parade ground at Fort Shafter, Hawaii, October 20, 1940. General Electric Corporation.

terrorize and demoralize civilians, destroying an enemy's will to fight. One might say that he invented the idea that air war should shock and awe the enemy. In 1937, with such predictions in mind, American military planners put the B-17 bomber into production.[15] By then, Douhet's theories were being tried out in Japan's bombing of China, in the bombing of Guernica, and in Italy's invasion of Ethiopia. Only in sober reflection and comparative study after 1945 did it become clear that aerial bombing of civilians usually did not break their will but angered them and increased their resolution.[16]

National responses to the new threat varied considerably. The British government attempted a blackout of London in 1939, but the public did not cooperate sufficiently and the city was by no means hidden.[17] The French held similar exercises; perhaps to win over the populace, they focused much attention on developing tasteful alternatives to black air raid curtains. Their studies found that "blackout effects can be achieved without eliminating all light and color." A decorator could use layers of red and green glass, "as red absorbs everything else and the green in turn absorbs the red rays. A combination of blue and orange is equally effective. These facts lead to possibilities for curtain linings, colored shades, etc., for which complementary colors could be used."[18] The matter was left to private preference, but among the general population "these scientific facts do not appear to have been widely recognized." (In any case, the French surrendered so quickly that Germany occupied Paris without bombing it.)

The Germans, less concerned with aesthetic fine points, used black curtains only and were relentlessly effective. In 1937 their war preparations included a blackout of Berlin that tested various forms of air raid curtains and shades on traffic lights, as well as trying out powerful searchlights that combed the sky looking for attacking planes. The Third Reich demanded that Berlin's 4 million inhabitants observe the blackout for an entire week. Although restaurants and businesses could remain open after dark, their windows had to be shrouded. Traffic lights were turned off, flashlights were *verboten*, and a person could be arrested merely for striking a match. Precautions became even stricter whenever an air raid punctuated the "black week." Then all traffic, including trams and local trains, came to a standstill, and people had to take shelter underground. Because the theatre of imaginary battle extended well beyond the city, "methodical preparations" were said to have been made for "the general blackout in every city, town, and village in the countryside."[19]

Pacifists worried that practicing blackouts prepared the population for war and trained civilians to accept military command. There seems to be little evidence that many Americans protested the blackouts. But in 1938 a group of Danish pacifists in Roskilde decided to sabotage three days of blackouts in Copenhagen. They purchased fireworks, which they planned to set off during the blackout "to show the whole world that not everybody wants to cooperate in preparations for the next world war."[20]

In a sense, the Danish pacifists were correct, for the purpose of a blackout was not merely defensive. In Berlin, a "sham battle" was waged over the darkened city, with anti-aircraft batteries "firing" at planes whose pilots tried to hit pre-arranged targets. In each military exercise, air squadrons flew overhead during drills, dropping flares, taking photographs, trying to "read" the blackout, and then testing accuracy by dropping harmless dummy bombs. In the United States, such exercises began in May 1938. Only a few areas participated, and cities that only heard the planes overhead, such as Hartford, were a bit disappointed to be left out.[21] The "attack" focused on Long Island, where the town of Farmingdale and its surroundings held a 30-minute blackout. Not only was the time short, but the area darkened was a small circle, two and half miles in radius. Because the military assumed that an actual raid would not bomb civilian targets, attention focused on protecting an aircraft factory. "This spectacular 'blackout' and 'bombing,'" the *New York Times* noted, "will climax the war game that has roared all along the Eastern seaboard. . . ."[22] Although the streetlights were off, from the air the movements of automobiles traced out the street patterns, making it easy to locate the target. Furthermore, the raid showed the effectiveness of flares. After the lights were extinguished at 22:32, "suddenly, the whole sky was ablaze from flares dropped by observation planes locating the target for bombers at 15,000 feet."[23] They were so bright that "they paled a full moon, and with their light it was possible to pick out the main highways, the Seversky aircraft plant and other vulnerable points." Impressive as the exercise may have seemed to the thousands who came out to watch, in comparison with the blackouts in Tokyo and Berlin it was a small-scale event.

In October of the same year, the American military held a more extensive drill in the Carolinas, near Fort Bragg, which involved 66 towns and 750,000 people. Two thousand civilian volunteers worked as spotters, seeking to locate the invading aircraft that attacked an open field that the exercise designated as a major military base. The army brought all of the 24 anti-aircraft weapons east of the Rockies to the site.[24] The organizers discovered a discouraging paradox: "On the ground the blackout, particularly in the Fort Bragg area and in the vicinity of Fayetteville, a town of about 15,000, seemed highly successful," according to the account in *Time*. But from the skies, the landscape remained quite legible. "From the air, the lights of towns, roads, and automobiles were clearly visible and the raiding bombers had no difficulty finding their objective. . . . Inability to darken scattered rural homes and keep cars off highways in so large an area defeated the blackout. Bombers found their way with ease, [and] theoretically wrecked Fort Bragg."[25] The Army discovered that blackouts had to be carefully organized and absolute in order to work. It also found that its new 800-million-candlepower searchlights could not penetrate clouds at 6,000 feet and could not pinpoint the attacking aircraft at 10,000 feet. At best the planes "were only 'flicked' by the lights."[26] From the point of view of those on the ground, the military blackout created an obscure but still navigable landscape. Electrical service had not been cut off, and normal activities could continue. But from the air, even without using flares, the faint visibility of towns and roads marked them for destruction. Traces of electric light, particularly from automobile headlights, made a community vulnerable. This realization came rather late to the British. Just weeks before World War II, southeastern England readied itself for a major drill. American newspapers reported that on August 11, 1939, at 12:30 a.m., 25,000 square miles of England was plunged into total darkness, while 800 "enemy" bombers "roared in from the eastern and southern coasts to see if they could penetrate the defenses of 800 planes and 60,000 air defense soldiers."[27] In fact, British pilots found "blacked out" areas easily visible. "Had the war broken out suddenly yesterday it is now apparent that London would have been caught woefully unprepared and, in the words of one observer, 'blown to smithereens.'"[28] It was more difficult to locate some sites at dusk than three hours later, when even faint illumination painted a picture of the targets below. But perhaps the most worrisome discovery of this exercise was how easily invading bombers could evade detection. One bomber crew "dodged all the searchlights—even over the naval base at Chatham—and passed unchallenged—once right over a fighting plane looking for them—until they reached their objective . . . almost in the center of England."[29] The plane was then able to fly undetected back to the Netherlands and return to its British base "not by dead-reckoning but by the lights that should not have been showing."[30] It was considerably easier to hide the lights on an attacking plane than it was to darken a city on the ground. In view of the difficulties of the task, it was fatally unfortunate that the civil authorities coordinating blackouts often lacked authority to compel compliance, and had to rely on voluntary citizen cooperation. One result of the failed London blackout of 1939 was to give them more powers.

Once the war had begun, the British took the threat more seriously. Housewives darkened their windows and in the process "exhausted stocks of black paper," dark cloth, thumbtacks, and cellophane. "Workmen hooded street signs, to prevent the reflection of lights, placed black screens broken only by an 'X' over traffic lights, and shaded lamps in subway stations."[31] Many children were sent to live in the countryside. Business files were moved to safe locations, and sandbags were piled around strategic sites. Luckily for the British, a sustained German air attack did not begin for almost a year.

During the war, blackouts became a way of life for civilians on both sides. Paradoxically, while production and use of electricity increased markedly, its visible consumption fell just as dramatically. To someone in the darkened streets it may have appeared that power was in short supply, but in fact every country involved in the conflict increased electricity generation. In the United States, production jumped from 161,000 gigawatts in 1939 to 271,000 GW in 1945.[32] In the same years, Great Britain's output rose from 26,000 to 40,000 GW, while Germany's went from 61,000 GW in 1939 to 74,000 GW in 1944, after which time the records are fragmentary.[33] Even coalless Italy, relying almost entirely on hydroelectric power, surpassed its 1939 production levels by at least 6 percent in each of the first four years of the war, until generation fell 30 percent in 1944 during the disruption of defeat.[34]

All combatants increased power production despite bombing runs on power plants. (See figure 2.3.) By the fall of 1941, experts had studied the experience of attacks on the electrical system in the warring countries. "In the case of widespread

Figure 2.3
A daylight bombing attack on a Cologne power station. Library of
Congress.

storms," they concluded, "the damage to be dealt with has been
vastly more extensive than that which bombing occasions."[35]
An ice storm or a hurricane was comprehensively destructive,
with the potential to rip down every overhead transmission
wire; by comparison, the effects of bombing were small-scale
and random. Underground cables proved quite durable even
when bombs fell near them. Surprisingly, overhead lines weren't
much more susceptible to damage.[36] When generating stations
themselves were directly hit, often they were not "put out
of action completely or permanently as a result of enemy
bombing."[37] In part, this was because modern generating
stations were fireproof buildings.[38] However, bombing often

disrupted electricity supply indirectly by starting fires and by damaging the water-supply system.

Londoners' heroic efforts to maintain a semblance of ordinary life while blacked out and under bombardment became legendary. Defying air attacks, thousands of men and women in the gas and electrical industries daily risked their lives to maintain essential services.[39] After nine months, no power stations had been permanently wrecked, in good part because blast walls had been built to protect them. Thirteen depots established around London held reserves of spare parts, so that the transmission system could be repaired rapidly.[40] Power was seldom disrupted for long, and restaurants, movie theaters, hotels, and other public buildings remained open, though they had to install a "light lock" at the entrance, consisting of a darkened passage with at least one corner to prevent light from spilling into the street.[41] In his 1940 radio broadcasts to America, Edward R. Murrow emphasized that crowds still gathered at the Savoy Hotel, where the French chef still managed to serve excellent food despite rationing. The blackout did not prevent dancing in the ballroom. But listeners also heard descriptions of the darkened streets, the air raid sirens, the firing of anti-aircraft guns, and bombs exploding. They learned that German planes had been spotted at the coast, heading for London. Once the warning was telephoned in, ambulances, hospitals, and doctors prepared to deal with the wounded. Until sirens sounded, buses continued to move, but they ran almost completely dark. Public lighting was permanently extinguished throughout the war while powerful searchlights crisscrossed the heavens searching for German planes. Crowds descended into the subway systems and underground shelters (figure 2.4). Through his broadcasts to the United States, Murrow taught Americans the wartime

Figure 2.4
Londoners camp out in the underground during a German air raid.
Associated Press.

necessity for vigilance, organization, stoicism, humor, and
courage as exemplified by the British.[42] This was a somewhat
idealized picture. More than 100,000 homes were destroyed in
the war, and 20,000 Londoners died in the Blitz as Germans
dropped 20,000 tons of bombs. The death toll was considerably
less than had been projected, however. The Air Ministry had
estimated more than 18,000 deaths per week of aerial bombardment.[43] Yet less than 5 percent of the population spent the night
in London's subways. Most chose instead to remain at home,
hiding under specially designed iron or steel tables or in bomb
shelters dug in gardens. Strategic bombing proved less accurate
than had been expected, even when the Royal Air Force did not
drive attacking bombers off target.

Solidarity in the face of danger was by no means universal. In London, crime flourished under cover of the blackout, which made police work difficult.[44] German bombs ripped open houses and shops to thieves, and shortages increased the value of second-hand items. Houses that suffered direct hits usually had to be demolished, and demolition workers sometimes stole. "It's the funniest bomb I ever come across," a foreman ironically declared. "I been all through the last war and I done several jobs in this, but I never come across a bomb like it. It's blown every bag open and knocked the money out, it's even knocked the money out of the gas meters, yet it didn't break the electric light bulb in the basement."[45] In the midst of the Blitz, the black market flourished. Crime increased as more than 10,000 deserters joined the underworld of vagrants and lawbreakers. Occasionally people returned from air raid shelters to find their houses unscathed but burgled. However, the murder rate did not go up, though a killer dubbed the "blackout ripper" briefly stalked London, killing four women before he was caught and hanged. From the criminal point of view, the war created demand, while the blackout eased access to supplies. Artificial darkness may have protected the city from accurate bombing, but it also hid the predations of looters and thieves. During the war, London was once again the "city of dreadful night" that Victorians had feared.

In the first months after the United States entered the war, blackouts and "dim-outs" were widely adopted and enforced until it gradually became clear that there was little need for them in most parts of the country. At first, however, most Americans worried about air attacks. On December 11, 1941, four days after the attack on Pearl Harbor, unidentified planes were reported over Los Angeles. Air wardens, police, firemen, and Boy Scouts sprang into action, and "much more rapidly than would have been believed possible three days ago, the lights blinked out over the whole city, until at 9 o'clock it lay almost lightless under a clear, starlight sky."[46] A Westinghouse executive warned that German bombers based in Norway could fly 7,000-mile round trips, which made blackout precautions advisable in Boston, Buffalo, Cleveland, Detroit, and Chicago.[47] Woody Guthrie wrote a song, "Gonna Be a Blackout Tonight," that urged people to put up blackout curtains and show no light.[48] Even remote towns in the Iron Range of northern Minnesota practiced total blackouts, enforced by monitoring from many volunteers. German or Japanese planes never did reach the Midwest, but civil defense preparations there were vigorous. Anyone failing to extinguish lights during a drill was likely to be dragged into a court, lectured by the judge, and fined.[49] Such exercises seemed more urgent and necessary along the Eastern Seaboard and parts of the West Coast.[50] Every street-light and every traffic light had to be shielded. In New York City alone, several thousand electrical workers volunteered to adjust the 28,000 control boxes so that air raid wardens could turn lights off by hand if necessary.[51] The Office of Civilian Defense systematized such local initiatives in national blackout guide-lines issued in March 1942. Families were made acutely aware of the danger as children were issued identification tags (similar to those worn by soldiers) to wear around their necks. These tags, which gave the child's name, birth date, and religion in case of injury or death in a raid, were constant reminders of danger. After making such preparations and holding air raid drills, people "assumed that the Germans would attack us."[52] A boy in Wisconsin dreaded blackouts, during which he sensed enemy planes "in the darkness overhead" and "thought about being bombed."[53] Air raids were not the only reasons to darken cities. German submarines operating along the East Coast often could see merchant ships silhouetted against lights on shore, notably at Atlantic City.[54] Before blackouts were put into effect, even the lights of parkways and buildings made ships visible as far as 40 miles out to sea.[55] Every sunken ship was both a physical loss and a graphic reminder that the enemy lurked nearby. Figure 2.5 shows two views of New York City during a blackout trial in March 1942. As late as November 1943, New York held an air raid drill, darkening the streets and extinguishing shop windows. Times Square went completely dark.[56]

Figure 2.5
The New York skyline from across the Hudson River before and during a blackout trial in March of 1942. Library of Congress.

In Britain and Germany blackouts remained essentially attempts at self-preservation. However, as the war progressed, bombing practices changed. During the first two years, the air forces on both sides expected that bombing raids could be precise, hitting railways, airfields, warships, munitions factories, and other military targets. Indeed, both sides gave assurances, either directly or through intermediaries, that they would not bomb civilian targets. However, accuracy proved an elusive goal, while intensive anti-aircraft fire put a premium on dropping bombs quickly. Some civilians were bombed in haste or in error. However, each air force gradually shifted its tactics toward destruction of civilian targets and to night bombing, which was safer for pilots and crewmembers. The result was extensive urban bombing that destroyed more than 800,000 houses in Britain. Yet the worst destruction occurred in German cities during the last years of the conflict.[57] Whereas an estimated 40,000 British died during the Blitz, 130,000 died in the bombing of Dresden alone. As Jörn Friedrich has documented, the firestorms that consumed Hamburg, Dresden, Frankfurt, and other cities were not accidental by-products of the war.[58] The blackout drills had shown in the 1930s that flares made cities quite visible despite their self-imposed darkness, and bomber crews often could see their targets even without them. The firestorms were intentional. In order to set entire cities ablaze, the Allies developed special bombs designed to start fires that lasted at least eight minutes after impact, long enough to ignite anything flammable nearby. The blackout as a tactic had assumed that the enemy would bomb selectively. It was not a particularly effective defense against wholesale destruction.

Once it became clear that German planes were not about to attack the United States, the government enforced "brownouts" to save coal. The brownout, a form of rationing first used in Australia in 1941,[59] was quickly adopted in America "to denote semi-darkening a city as distinguished from the complete darkening of a blackout." In 1943 Washington planned a "'brownout' (midway between total darkness and every marquee ablaze)" that "would help save electricity and fuel."[60] The War Production Board banned all daytime advertising lighting and permitted night advertising and the illumination of show windows only for businesses that were open, and only for two hours between dusk and 10 p.m. In addition, wattage for all signs had to be reduced as much as possible. Ornamental lighting and display advertising on Broadway was banned, though for two weeks New York's twenty-one official Christmas trees were permitted to be lit until 10 p.m.[61] Americans were also asked to eliminate waste at home, turning off lights and appliances when not absolutely needed.[62] Some flouted the brownout and were fined, but most took it seriously.[63] In January 1945 the War Production Board briefly re-imposed the blackout, prohibiting all outdoor advertising, ornamental lighting, and show-window illumination.[64] Entertainment was little affected by these regulations, however. Theaters remained open, even if their marquees were subdued. Night baseball was exempted from the regulations, since fans would not be consuming electricity at home.[65] Traffic accidents were more common during the blackout, however, and night traffic fatalities increased 51 percent in the early months of the war.[66]

As the conflict dragged on, people longed for the time when they could put the lights back on again. The British singer Vera Lynn caught this mood in a song titled "When the Lights Go

On Again," first broadcast in November 1942.[67] The lyrics linked the rekindling of full electric lighting to peace, the return of soldiers, freedom of movement, free hearts, love, and marriage. In peacetime, the only thing falling from the skies would be rain and snow, and a kiss would no longer symbolize a long separation or permanent loss. Lynn's lyrics longed for the time when the lights came back on and there would be time for weddings and "free hearts will sing." It was broadcast frequently during the war. Another popular song for two years before the war's end was "I'm Gonna Git Lit Up When the Lights Go Up in London."[68] Both songs linked lighting to a better, more peaceful world.

Yet in April 1945, when the end of the war was near, lifting blackout restrictions in Britain did not immediately elicit spectacular lighting displays. Newspaper reporters, who had expected to see a dramatic contrast to the previous 2,061 nights of darkness, were surprised to find little immediate change. The British people seemed accustomed to the gloom, and some said they felt exposed in a room without blackout curtains drawn. Even if one wanted to illuminate a shop, often the neon bulbs had been destroyed or packed away, in either case requiring an electrician, who generally was not available. After years of bombings, "not one store in a thousand has its exterior electrical equipment working."[69] The transition back to peacetime illumination would take time.

In New York City, where the blackout had become a three-year brownout, most lighting equipment had not been removed or damaged. The city burst into light at the first opportunity. "Fingers of light groped skyward as dusk fell last night, turning the semi-darkness that had been Times Square into an abyss of shimmering whiteness, while out in New York Harbor the Statue of Liberty, for the first time since early 1942, threw the light of its blazing torch into the bay." In fact, the 20,000-watt lamp was twice as bright as the one used before the war. In Manhattan, an enormous crowd blocked Broadway, staring up at the advertising signs, and reveling in the city's return to normality. One man put it this way: "Boy, New York looks like home again."[70]

Wartime blackouts completed the normalization of the electric landscape, which was identified as a natural state, in contrast to the dark abnormality of war. An illuminated environment had become the emblem of peace. Darkness, a familiar part of human existence for millennia, had become dangerous, artificial, and intolerable. The reappearance of public electrification after the war gave children who had grown up during the blackout their first experience of the electrical sublime as the glowing night city suddenly emerged in full regalia. A man from Leicester recalled such an experience: "My parents took me for a late walk on the night after the war had ended to show me the lights. The street lamps were on and shining in all of their glory, no house had black curtains at the window and it was wonderful to see all of the different patterns of curtain hanging and being lit up by the indoor light and the few cars that were about, all had their headlamps full on, it was like being in wonderland to see so many lights after I had only experienced nights of darkness. . . ."[71] In central London, the illumination was far more extravagant on the night after victory was declared than it had been 10 days earlier. "The night scene was consummated by the exhilaration of full-powered, aggressively-sported, lighting. The statues and public buildings were floodlit; the searchlights danced a ballet in the sky; from restaurants and cinemas poured a glow which seemed unearthly to the small children whose parents let them stay up late to wander in this unimagined fairyland of illumination."[72] In Paris, Armistice Day was celebrated by what Le Corbusier recalled as an "undreamed of revelation: a floodlit Place de la Concorde. Not just lit up by its street lamps, or the Republic's standardized little gas flames, but illuminated with all the floods of light made possible by electricity. The idea had come from America, the projectors from the war."[73]

Both before and after World War II there was a second form of intentional blackout. In some labor conflicts, unions used fuel shortages and power outages as a threat or a weapon. This occurred in industrial countries at various times, notably in the United Kingdom during the 1970s, but it was particularly widespread at the end of World War II, when many workers felt that they had boosted productivity without receiving corresponding wage increases. It seemed time to get deferred raises. In the United States, which entered the war two years later than Europe, union blackouts also occurred between 1939 and the attack on Pearl Harbor. For example, in New York the Electrical Workers Union shut down the lights of Times Square and Broadway in 1941 during a dispute with Consolidated Edison. For half an hour "the fish didn't swim in the Wrigley sign in Times Square; Fred Astaire and Rita Hayworth vanished from the illuminated square . . . , and the coffee in the magnificent Silex bubbled no more, though steam forlornly hissed from it." A crowd gathered to view the darkened scene, as more than 300,000 bulbs and 200,000 feet of neon tubing went out.[74] Not all utility strikes were merely symbolic. A few weeks later, in Kansas City, a sudden midnight walkout by American Federation of Labor workers caused a four-hour blackout that affected 400,000 persons as it halted streetcars, cut off the water supply, froze production in defense plants, and forced hospitals to operate by flashlight. The union was protesting against a company union that persisted, even after the National Labor Relations Board had ordered it disbanded. The blackout without warning angered the entire city, and the chief of police declared that if anyone died because of it he would charge the organizers with murder.[75]

As the war came to an end, conflicts between unions and management resumed. In New Jersey, in November 1945, the International Brotherhood of Electrical Workers threatened a blackout that was only narrowly avoided.[76] In addition to strikes at utilities themselves, nationally 200,000 electrical workers went on strike on January 15, 1946. Pittsburgh, home of the Westinghouse Corporation, was one of the centers of the dispute, with more than 18,000 electrical workers out. Neither of these strikes affected the local power supply, but at 4 a.m. on February 12 workers at the Duquesne Light and Power Company also struck, plunging the city into darkness. Pittsburgh's new mayor had appealed to the union to continue negotiations. In an emotional radio address, he declared: "I cannot believe that people whom we know, people who are part of our community, people who are one with us, will plunge our city into darkness—risk the lives of people near death's door in our hospitals."[77] But the city was plunged into darkness more than once in the ensuring months, as the power company refused to grant a 37 percent salary increase but came to a temporary agreement. Another round of acrimonious negotiations began in the autumn of 1946, and on September 25 Pittsburgh was again in darkness. "Night travelers to Pittsburgh had the unnerving experience of coming into a city illuminated largely by candles. And the farther they went into the dimmed out city, the deeper grew the sense of having strayed into some darker century."[78]

Streetcar service was cut by half, and most businesses closed. The city barely functioned. Office workers had to walk or hitch a ride into town and climb the long stairs to their offices, which had no steam heat or light. Most industries ground to a halt. All department stores closed for several weeks until the dispute was settled on October 20. By that time Pittsburgh was near collapse, and pressure on the union and the utility was intense.

Strikes persisted in many industries through the rest of the year, including a nationwide strike by coal miners. Though the public often assumed that most electricity came from hydroelectric dams, that was never the case. No major city in the flat Middle West has direct access to much waterpower, and, like Chicago, they relied primarily on coal. Despite the generators at Niagara Falls, most of New York City's power also came from coal-fired steam turbines. In 1946 coal supplied half of America's energy, not just for electricity but also for making steel, heating buildings, and myriad other activities. However, hydroelectric dams were still significant. In 1939 falling water generated 30 percent of US electricity, and construction of new dams increased capacity. Yet steam turbines were installed just as rapidly.[79] Even in the Tennessee Valley, the government dams were not long sufficient to provide enough power to the region. Starting in 1948, the TVA built conventional power stations. At first this was done only to supplement its dams, but by the 1960s the majority of the TVA's energy came from coal.[80]

City leaders took drastic measures during the coal strike of 1946. Fully aware of his city's dependence on coal, New York's mayor declared a state of emergency in February, when striking tugboat workers refused to move barges of coal over from New Jersey.[81] For a week the lights darkened in Times Square, but this was only a rehearsal for the coal shortages later in the year.

Miners of bituminous ("soft") coal went on strike April 1, followed a month later by miners of the cleaner-burning anthracite coal. By the end of April, the US had only enough fuel to last three weeks. All large cities felt the pinch, Boston, Detroit, Philadelphia, and Chicago worse than most.[82] On May 2 New York officials warned of possible brownouts,[83] which were widespread across the country a week later. A survey of midtown Manhattan at the end of May found that illumination had been cut in half.[84] Indiana and Illinois drastically curtailed electric power. Commercial customers were restricted to 24 hours' service between Monday and Friday, and department stores opened only from 2 to 4 p.m. All ornamental lighting was banned. In Chicago, utilities cut back 40 percent, and the coal shortage forced a widespread closedown of both industry and commerce. More than half of all industrial workers were laid off, movie theaters closed, shopping was limited to afternoons, and more than fifty long-distance trains were cancelled.[85] In early June, coal supplies improved after the strike ended. However, in the autumn a second coal strike forced a brownout that started November 25 and lasted for two weeks in New York, Chicago, Philadelphia, and other major cities.[86] Whole industries came to a standstill. The Ford Motor Company laid off 45,000 workers. As railroads ran short of coal, they began to cut back service.[87] Their problem was almost compounded by a threatened national railroad strike that was barely averted after some hard-nosed negotiating from President Truman, who threatened to call out the Army to run the trains.

During the second coal strike of 1946, many institutions, including Congress, voluntarily reduced lighting. After the architect of the Capitol turned off its floodlights, a group of American historians declared: "The dome of the Capitol is a cherished symbol to all Americans. Except for purposes of defense against an external enemy it should never be blacked out." Even during the Civil War, they noted, President Lincoln had insisted that workers continue the construction of the Capitol "as a symbol of the continuity and permanence of the United States." Keeping it lighted at night was necessary for the same reason. "And now are we going to darken it for [union leader] John L. Lewis?"[88]

Darkness was descending all over the United States in the autumn of 1946. Two weeks before Christmas, "New York browned out by the coal strike looked very much like New York blacked out by the war." The only difference was that private residences shed some light into the streets. But the scintillating signs of Broadway were dark. "Fifth Avenue Shop windows no longer brightened the way," and generally the streets were "caverns of gloom." Aside from the discomfort of the darkness itself, the psychology was also different. The "wartime blackouts expressed a sense, not so much of imminent danger as of alert public responsibility." But to most New Yorkers, whether they supported it or not, the coal strike emphasized not unity but class division. During the war, "there was a sort of comfort in that dimness. Our glittering, glaring city took on a surprising soft loveliness." In contrast, during the brownout, the city seemed "melancholy."[89]

Utility strikes also occurred during the fall of 1946 in Japan, where the occupying US forces found workers restive. During October and November, Tokyo's electrical workers demanding higher wages repeatedly cut off power for 5 minutes at noon.[90] In Japan's second-largest city, Osaka, streetcars, buses, and subways were halted in a 24-hour strike for the right to collective bargaining and closed shops.[91] Even though the 95,000 electrical workers were better paid than many others,[92] they managed to win after a fifty-day dispute. A four-hour nation-wide blackout was averted when the government raised their wages to $1,600 a year.[93]

In France in the fall of 1946, coal shortages forced electrical blackouts for all customers two days a week on a rotating basis. Blackouts were also necessary in occupied Germany.[94] In Britain during February 1947, a period of unusually cold, stormy, and snowy weather closed many rail lines.[95] Storms prevented coal ships from sailing, and trains were unable to run because snow blocked the tracks. Some villages were cut off for ten days or more, and the Royal Air Force had to airlift food supplies to them. Power plants ceased receiving coal, and as their stocks dwindled to a two-week supply (vs. the usual six-week supply) Clement Atlee's Labour government had to impose a blackout as an economic measure. The streets of London, Birmingham, and Manchester were dark again. Pubs remained open but used candlelight. Many offices relied on lanterns. Millions of domestic electricity consumers were cut off, and the rest were asked to curtail consumption. Factories closed, and millions lined up for the dole. The crisis seemed almost as severe as the war itself, and underscored how essential energy had become to the functioning of a modern society.

There were several causes of the British coal shortage of 1947. The Labour government had nationalized the coal industry, introducing the legislation in 1945 and passing it in June 1946.[96] Nationalization was intended to rationalize the industry, but the transition was awkward. Private mine owners about to lose their holdings had no incentive to modernize or to invest in new equipment. Some skilled managers left mining for other employment. Aside from the organizational problems of a transition to public ownership, there were technical issues. Though Britain had reserves good for 200 years, coal extraction had been under way since Roman times. As mines burrowed deeper toward less accessible seams, productivity per worker declined despite the mechanization of coal cutting and hauling. Furthermore, the industry was severely undermanned at the end of war, and the average age of miners was high. Many had delayed retirement to do their part in the war, but stopped working when it ended. By 1947 there was an acute need for 100,000 additional miners. Sons who in peacetime might have taken their father's places had gone to war. Some had died or been wounded; others had learned new trades. Nor was mining attractive to outsiders. It was dirty, dangerous, poorly paid, and uncertain, with a volatile history of layoffs. Irregularities in the coal supply made other energy supplies attractive to consumers. In the United States, coal supplied half the energy during World War II, compared to 44 percent for oil and gas. But between 1945 and 1960 coal's share fell to a quarter, while oil and natural gas use shot up to 73 percent of all energy consumption.[97] The ability of striking coal miners to disrupt electrical service declined correspondingly.

In retrospect, the most salient feature of the blackouts that occurred between 1935 and 1947 is technological control. The early military blackouts were conscious impositions of administration in order to create a comprehensive special effect. From 1880 until the early 1930s, electricity had been used not only for practical purposes but also for extravagant display. Then the mastery of flight—itself a form of the technological sublime and a thrilling spectacle—demanded, from the military point of view, the imposition of darkness as a security measure. The simplest way to achieve this would have been to shut down the power stations. But although this might hide a city from a night bombing raid, it would also paralyze it. Blocking out the lights was far more desirable. Thus, there was a double movement in the mastery of electricity. First came the desire for spectacle, extravagantly realized in Times Square and Piccadilly Circus. This first movement celebrated modernity, demonstrated mastery over nature, and encouraged utopian visions of the future. But next came the subordination of the electric landscape to the military's quest for strategic invisibility. The inversion of the electrical sublime was simulated darkness, punctuated by searchlights that probed the night sky. Yet the coal shortages and electrical strikes immediately after World War II demonstrated that blackouts could also result from cultural breakdowns, from bad weather, and from social failures.

In subsequent decades, the military blackout disappeared. The experience of the war had demonstrated that blackouts were most effective if assisted by bad weather, and that they often failed to fool the enemy. More to the point, blackouts lost importance as the military acquired the technologies of night vision and global positioning satellite systems. Once bombers could find targets in the dark, hiding the lights became pointless. After 1945, the blackout ceased to be a military necessity and became a civilian problem. In the second half of the twentieth century, blackouts would no longer express technological mastery and comprehensive social control. Rather, they would be accidents.

By 1950, ubiquitous power and light were thought of as "natural" and yet "civilized." Darkness was increasingly the realm of the unfamiliar, the strange, or the primitive. The naturalization of the electrified world led Americans to entertain further expectations. Once electricity was ubiquitous, instantaneous communication seemed a logical, perhaps even inevitable development in a line that ran from the telegraph and telephone through radio, television, the computer, and the comic-strip character Dick Tracy's two-way wrist radio to the mobile telephone. It also became "normal" for objects to interact across enormous ranges of space, linked into a system that monitored movements, tracked transactions, and left electronic "footprints." The first artificial satellites dramatized this development, and Americans became accustomed to the idea that scientists and engineers in a Houston control room could manipulate distant electrified objects. Once these communicative and cyber-kinetic possibilities were familiar, the electrical supply network seemed so ordinary and banal that it disappeared from most people's awareness.

As early as 1947, Americans began to call unexpected power outages "blackouts."[1] In the new usage, "blackout" meant far more than a plunge into darkness. It referred to a comprehensive failure of almost everything to function. It began to acquire the connotation of losing consciousness, taking on an almost physiological meaning. Like a knocked-out boxer, a city without electricity fell into suspended animation, unable to move or communicate. A blacked-out city, with security cameras and alarm systems not working, temporarily lost its nervous system. The new usage also owed something to the phrase "news blackout," which had become common during World War II. The more neutral term "power outage," which had fewer social and psychological connotations, gradually faded from use.[2]

In 1964, Lewis Mumford speculated that "something like catastrophe" had "become the condition for an effective education" that would demonstrate the public's dependence on the grid. "This might seem like a dismal and hopeless conclusion," Mumford wrote, "were it not for the fact that the power system, through its own overwhelming achievements, has proved expert in creating breakdowns and catastrophes."[3] Mumford knew that New York City had the largest and most complex electrical system in the United States. Despite Consolidated Edison's self-interest in maintaining a reliable system, it suffered periodic breakdowns. Mumford recalled that in 1959 and again in 1961 several square miles of Manhattan had been blacked out. In 1961, during an early-summer heat wave, as the thermometer reached 96°F, the area between Forty-Third Street and Seventy-Seventh Street lost power for more than four hours.[4] Elevators and subways trapped sweltering workers about to go home. Without signals, traffic became snarled.[5] These were portents. They were not caused by shortages of electricity itself, but by failures in the infrastructure that delivered it. This included not only malfunctioning generators and switching equipment but also the monitoring and control systems, which were becoming ever more automatic and instantaneous. As operators became responsible for larger networks, it was becoming difficult to interpret and use information as rapidly as it was received. When lightning struck a substation, when a fire broke out in a power plant, or when a high-tension line shorted out, the loss of the regular flow of electricity in one part of the regional grid had to be made up from some other source in a just few moments, preferably in just a few seconds.

Even as controlling the production and distribution of electricity became more complex, customers depended more on the system. Each overlay of new electrical devices increased the disruption that a blackout could cause. And yet, although a blackout became an ever greater, more expensive, and more comprehensive disruption, few outside the utilities thought much about it until 1965, when the largest power failure in history struck the northeastern United States and parts of Canada. It darkened New York City, southern New England, and the area to the west all the way to Toronto. At that moment, the blackout attained entirely new meanings. By 1965, neither the blackout as minor disruption (c. 1920) nor the blackout as wartime necessity (c. 1939) made phenomenological sense. A comprehensive blackout was a powerful form of negation, not merely extinguishing the lights but disrupting all the functions of society. And few Americans under the age of 40 had practical experience in coping with an unelectrified world.

In the 1930s, a blackout was an intentional military erasure of peaceful modernity; later in the century, its primary meaning would be that of unintended and unexpected failure. It was no longer a temporary, planned response to an external threat, but rather an unplanned disaster of indeterminate length, an irruption of systemic weakness, a form of irrationality that came from within. The military blackout was a comprehensive scheme of camouflage imposed upon the underlying electrical system, demonstrating mastery and efficiency. In contrast, after 1965 a blackout demonstrated incompetence or malfunction. It did not hide an underlying order and purpose; it exposed an underlying disorder. It paralyzed the infrastructure and revealed technical failures and management miscalculations.

The unplanned utility blackout of 1965 struck without warning, disorienting citizens. Yet during a power failure a city was not entirely dark. Indeed, in some locations it was brighter than it had been before the electrical system began to be installed in 1880. During a "blackout" the lights of automobiles clearly marked the major highways, and the backup emergency lighting in hospitals, police stations, and some public buildings provided orientation, supplemented by the autonomous systems of fire trucks and ships. Yet a power failure stranded people wherever they happened to be, shut off ventilation and air conditioning, caused computers to crash, and closed down almost every business. A military strike that achieved half as much would be judged a brilliant success.

The effects of a large power failure were far less serious before World War II. In the 1930s, dependence on electricity was still recent and not yet entirely normalized. Local electrical systems often broke down, and people took outages in stride. Power losses did not yet seem to be extraordinary ruptures in experience.[6] As electrical utilities merged into regional systems, however, the potential for widespread power failure increased, although the social consequences were not yet dire. Fewer activities depended on electricity in 1935 than in 1965. For example, in January 1936, after a short-circuit at New York Edison's Hell Gate generating station, the power went off in the Bronx, in Manhattan above Fifty-Ninth Street, and in much of Westchester County. This occurred at 4:16 p.m., near the peak of demand. "Lights in offices, hospitals, department stores, homes, and industrial establishments throughout the area began to flicker and then failed completely."[7] Traffic lights went black, movies stopped, elevators stalled between floors, and the subways halted, stranding 60,000 people. When power began to come back on, around 5 p.m., the current was weak, and the subway switching system did not work properly. Cars could not climb grades on some parts of the line. As traffic lights began to work, some stayed permanently on red, others on green. Life did not return to normal for three hours. Yet the fact that Brooklyn, Queens, Staten Island, and the lower half of Manhattan were not affected is a reminder that the electrical system was not as tightly woven together as it would be by 1965. In 1936, a blackout that cascaded from one region to the next was still rare, and it was not only possible but likely that just half of New York City would lose power.

Temporary power failures had been quite common after 1882, when Edison's electrical system was installed in New York's financial district. At the end of 1883 that system had only 500 customers, and they did not expect the power to be available without interruptions. When generation problems emerged, America's first power station had nowhere else to turn to get electrical current, and the system went down. Expectations of continuous service developed only gradually. In 1901 the local grid was knitted together as the result of mergers between many small power stations, creating Consolidated Gas (later Consolidated Edison). Occasional outages long seemed "normal," and people kept candles, flashlights, and other backup technologies ready to hand. In the 1936 blackout, New Yorkers grumbled a bit but expected that service would soon be restored. The same was true of the citizens of Newark, New Jersey, who experienced a 5-hour blackout in January 1936, again in the late afternoon. The outage resulted in only temporary difficulties and only one traffic accident. Office workers climbed down ten, twenty, or thirty stories and joined factory workers who had been sent home early. "All hurried away from the gloom-swept center of the city, where the shadowy streets [were] lighted only by the dim beams of a bright moon" (assisted by a few acetylene torches at intersections).[8] Bamberger's department store, which had its own generating system, stayed open.[9]

During the 1930s, Americans were not much disconcerted by temporary power outages. When a power system went out, most people still recalled everyday life before electricity. The vast majority of Americans had grown up without it. Even in 1936, most homes still had iceboxes rather than refrigerators, and very few had freezers. Many women had adopted electric irons, but they still knew how to heat up a solid metal iron on a stove (seldom an electric one). Offices too could get along without electric power. Typewriters and adding machines were mechanical. Copies of documents were produced using carbon paper and saved in file cabinets. A modern office, even as late as 1950, needed electricity primarily for light and ventilation. A power failure was a nuisance but not an insurmountable hurdle.

Consumption of electricity had to become more central to society before a blackout could become a landmark event, as first occurred in 1965. The Great Northeastern Blackout remained etched in popular memory a generation after it occurred, and not only because it affected a vast area. Just as important was the fact that the public had discovered many more uses for electricity in the two decades after World War II. When all places could be lighted or supplied with electricity equally well, it was not only possible but attractive to locate a factory, a store, or a home outside the old urban core.[10] The expanding network of electric lines encouraged a suburban diaspora. Work and commerce could have retained a tighter urban focus, but Americans chose to use electricity for an unprecedented urban deconcentration. Office buildings and the new shopping centers could be built anywhere with almost complete disregard for natural light. Enclosed malls had few windows to the outside world. Office towers grew wider, as corporations assumed that proximity to a window mattered less when employees enjoyed good artificial light and ventilation. Just as electrification once had been used to transform both the layouts and the locations of factories, it underlay the disaggregation of the walking city, the gradual invention of what came to be called "edge cities," and the expansion of suburban space.[11]

At the end of World War II there were fewer than a dozen shopping centers in the United States. By 1960 there were more than 3,800, and the enclosed mall, which first appeared outside Minneapolis in 1956, was emerging as the norm.[12] Americans zoned new, highly electrified spaces for single activities: suburban tracts, shopping malls, corporate estates, sprawling universities. The new form of living beyond the old urban core segmented life activities into distinct physical realms. There was nothing inherent in electricity that led to this result, for European cities did not deconcentrate to the same degree. It was a particularly American choice, and no institution expressed this more clearly than the suburban house.

The domestic consumer found many new uses for electricity in the decades after World War II. Houses changed after the introduction of wiring. Electricity permitted new forms of architecture and encouraged new kinds of social relations. Frank Lloyd Wright's open-plan architecture, in which rooms flowed from one to another, was an early harbinger of change. The Victorian house had fires in each room—stoves, fireplaces, candles, lanterns, and gaslight—which made windows and doors essential to control drafts and to air out fume-filled rooms. Houses with electrical lighting did not need to separate spaces. Doors were less needed to control airflow, and windows were no longer as necessary for ventilation and light. Likewise, dark color schemes, which concealed unavoidable traces of smoke on the walls from candles, gas jets, and lamps, could be abandoned for lighter colors. Electrified houses opened up and brightened. Their owners acquired vacuum cleaners, water heaters, and washing machines. Almost anything a cook could think of could be done electrically, and home economists encouraged housewives to buy toasters, coffee makers, waffle irons, popcorn poppers, mixers, grinders, choppers, refrigerators, ovens, grills, disposals, and freezers.[13] With power available at the flick of a switch, it was no longer necessary to haul wood and ashes or to clean lamps. Children, who could not be trusted alone with candles and lanterns, could turn on electric lights. Parents no longer needed to watch them as much, and families also had less need to cluster at night around the hearth, because warmth and light could be found anywhere in the home, which was also becoming larger. As families spread out, the privacy of each person increased, counterbalanced somewhat by more open floor plans.

As the grid expanded, Americans were encouraged to imagine that an electrified house could be just about anywhere. The New York World's Fair of 1964 depicted future homes in Antarctica, under the sea, and in the Amazon jungle.[14] Electricity equalized all spaces, delivering light, power, climate control, and information to any site, from outer space to underground caverns. But as electrification was used to homogenize space, it also made possible dense agglomerations of people. At night, airplane passengers saw patterns of lights that marked the settlement patterns, highways, and city centers, and in dark rural landscapes they also saw the depopulation of rural America that had occurred between c. 1920 and 1960. Satellite photographs of the United States revealed both the concentrations of energy and population in metropolitan areas and the darker, largely uninhabited areas around them.

Such continental transformations were hardly imagined in the 1870s when ordinary men and women explored electricity's possible meanings. Often the body was literally the contact zone between people and the new technology. Some doctors claimed to cure a wide range of illnesses and conditions by passing electrical current though the afflicted part of the body, and enthusiasm for electrical "cures" lasted well into the twentieth century. Despite the efforts of the American Medical Association to discredit them, people kept buying electrical "cures" until at least the 1930s.[15] Thousands bought electrical collars and belts for rejuvenation, because they regarded electricity as a magical fluid, a nerve-tingling "juice," or an invisible tonic that could restore normal health. From the 1870s until 1920, before many people had wired their homes, electrical medicine offered direct contact with the mysterious force that was transforming work and public life. Nor have consumers given up on using electrical devices to reshape their bodies. Some still buy electric belts advertised on television in the wee hours, and others attach electrodes to their stomachs with the expectation that they can effortlessly tone up their muscles and look thin. Millions also use private gyms with Stairmasters and other electrical devices that time workouts, monitor heartbeats, and count calories lost.

As battery technology improved, electricity was also used to enable the body in new ways. By the 1950s, tiny lightweight batteries could deliver small but constant amounts of power for a year or more, making practical electric watches, hearing aids, pacemakers, and other ingenious medical devices. Slightly larger batteries combined with new transistors made possible radios that could be carried in a shirt pocket. As batteries became lighter and lasted longer, they became assumed parts of everyday life. They drove electric toothbrushes, cordless drills, and other tools. They powered tiny portable radios, disc players, mobile telephones, iPods, and portable computers. They became essential to thousands of children's toys, even if they were not usually included when purchased. In outer space, satellites relied on solar-powered batteries to keep them operational for years. In little more than a century electrical devices had penetrated into the most intimate parts of the body and reached millions of miles into space.

In a parallel process, electricity was woven into the metaphors Americans used to describe themselves. Energy is not merely useful. Each successive form of power has been incorporated into self-awareness. The "human race horse" of 1820 became the "human locomotive" of 1860 and the "human dynamo" of 1920.[16] The popular adoption of metaphors in which wires, currents, and machines replaced muscle power marked a shift in consciousness that occurred along with the expanding use of electricity. By the 1930s at the latest, the American consumer conceived of the ideal self as "high voltage." Successful people were "powerhouses." Good musicians gave "electrifying performances." American speech became saturated with expressions that suggested electricity enlivened personality, quickened intelligence, and promoted growth. In the comics, a "bright idea" was symbolized by a light bulb. Alternately, energy shortages were translated into negative qualities. A stupid person might be called a "dim bulb." A confused person had gotten his "wires crossed" and suffered a "mental short circuit." A tired person needed to "recharge his batteries." Memory failure was called a "blackout," as was a temporary loss of consciousness. Because electricity provided characteristic metaphors for intelligence and consciousness, by almost logical extension more electricity surely was a good thing, quickening both the economic and the mental tempo.

The implications of electrical metaphors were made explicit in social-science theories that connected higher energy levels with higher levels of civilization. In the 1940s and the 1950s, the anthropologist Leslie White argued that "culture evolves as the amount of energy harnessed per capita per year is increased," and that "the degree of civilization of any epoch, people, or group of peoples, is measured by ability to utilize energy for human advancement or needs."[17] White posited four central stages in human history: hunting and gathering, agriculture, steam power, and the atomic age. His evolutionary typology valued most those cultures that controlled the largest amounts of energy. White adopted a position close to technological determinism: "Social systems are functions of technologies;

and philosophies express technological forces and reflect social systems. The technological factor is therefore the determinant of a cultural system as a whole."[18] During much of the Cold War, it seemed axiomatic that economic growth correlated with rising energy use. Yet, as cultures developed beyond muscle power, they became dependent upon its technological replacements. Any prolonged blackout was a serious threat not only to the economy but also to the whole social system.

During the Cold War, Americans understood the drab streets of Moscow to be inferior to the scintillating streets of New York, with its brilliantly lighted skyscrapers, huge electric advertisements, and well-stocked stores. At an exposition in Moscow in 1959, a modern kitchen stuffed with electrical appliances symbolized American energy abundance. This unexpectedly became the location of a debate between Richard Nixon and Nikita Khrushchev over the relative merits of communism and capitalism. As they confronted each other over a new automatic washing machine available in a selection of pastel colors, Nixon praised the American consumer's many choices between styles and manufacturers; Khrushchev replied that such multiplicity was not a sensible approach. Viewing an automated kitchen, Khrushchev sneered: "These are gadgets we will never adopt."[19] Implicit in the entire American exhibit was a limitless supply of electricity. Indeed, in the same years atomic power was being touted as a permanent source of electricity that would be "too cheap to meter."[20] Today, it appears that both Khrushchev and Nixon were wrong. The Soviet Union broken apart into many countries that abandoned communist ideology, and their citizens widely adopted the appliances Khrushchev ridiculed. Nixon's assumption of infinite energy abundance was also mistaken. Atomic energy costs more than most other forms of power, and social scientists no longer argue that energy use measures the level of a country's civilization. Yet the idea persists in popular culture, reinforcing the unspoken assumption that energy abundance is somehow the "natural" and "developed" condition, as opposed to the "undeveloped" and "dark" parts of the world.

To consumers in the Cold War period, an electrical utopia seemed imminent. Automation would usher in a pushbutton world. The work week would shrink, salaries would rise, and there would be an excess of leisure time. World's fairs in Brussels (1958), Seattle (1962), and New York (1964) trumpeted this vision.[21] In Seattle, for example, the General Electric pavilion displayed a home of the future equipped with "colored television projected onto large wall surfaces, an electronic home library, movies that can be shown immediately after they are filmed, a cool-wall pantry, pushbutton electric sink, electronic bakery drawer, clothes conditioning closet, and the home computer for record-keeping, shopping and check-writing."[22] Though such things were then considered utopian, in building codes and in the "war on poverty" electricity became a legal requirement akin to a natural right. Middle-class houses began to have not only kitchen appliances and televisions but also air conditioning, garden lighting, and garbage disposals. Before the rekindling of feminism, women were expected to be the "home managers" of this domestic utopia.[23] In contrast, during the early decades of the Cold War, Eastern Europeans had fewer appliances, and socialist governments experimented with shared appliance use and with motors that could be attached to several devices, minimizing duplication.[24]

In the United States, electrical lighting early moved far beyond any functional necessity and became a popular form of display.

By 1910, Times Square and the "Great White Way" had become national icons and had been copied to a degree by other American cities. Atlantic City, Coney Island, and other amusement areas likewise intensified illumination, in a process that eventually culminated in the hyperactive cityscape of Las Vegas.[25] At Christmas, homes were covered with flashing electrical lights, mimicking the aesthetics of the Great White Way. These commercial lighting forms were later refigured as art. Galleries decontextualized neon from the tawdry realm of advertising and made it highbrow.[26] Americans also used electricity to transform public events. Rock concerts amplified and intensified experience, as did sports stadiums that used loudspeakers, giant scoreboards, and instant-replay screens. Political parties adopted these technologies and transformed their national conventions into electrified media events. As Americans enjoyed this intensely electrified world, they naturalized its growth. Access to energy increasingly seemed an entitlement and an inalienable right. Because it literally empowered the individual, electricity became inseparable from a sense of physical well-being. Americans' expectations to do things electrically were based on habits acquired in energy-intensive homes, factories, offices, shopping centers, and leisure sites, and were reinforced by high-energy metaphors of the self.

Enmeshed in electrical metaphors and accustomed to consuming increasing amounts of energy, Americans confronted blackouts as temporary irruptions of irrationality. By the 1960s, the majority of Americans could no longer remember an unelectrified world. In one sense, all New Yorkers experienced in 1965 the same thing that half the city had experienced in 1936, when the lights went out and all electrical devices stopped functioning. Yet these two blackouts were apprehended very differently. As dependence on the system had increased, knowledge of how to live without electricity had declined. For example, someone writing at a desk in 1936 used either a manual typewriter or a pen. By 1965 someone writing at a desk probably used an electric typewriter; though it froze up when the power went off, no work was lost. But during the blackouts of 1977 and after, computer crashes erased millions of pages. In countless ways, New Yorkers realized during the last decades of the twentieth century that their city was a technical system of increasing complexity, woven together through electrical connections. With each passing decade, more elements of the city, when deprived of power, became paralyzed parts of a vast broken machine.

The stasis in a city suffering a blackout after c. 1960 could immediately be seen in the streets. Traffic snarled at first but gradually thinned out, while pedestrians became so numerous the sidewalks could not contain them. A blackout redefined the potential uses of public spaces and demanded improvisation. Public spaces are ordinarily designed for certain activities that are implicitly understood by adults, though children often invent "inappropriate" uses for them. For example, sidewalks are intended for pedestrians. If children sit down on a sidewalk, their behavior is considered inappropriate because it blocks the flow of foot traffic. But during an electrical blackout, many people sit on the sidewalk and put their feet in the gutter. Instead of hurrying along, they have nowhere to go. As the BBC reported, during the 2003 blackout in New York City there were "thousands of people lying on the sidewalks, propped up against shop fronts, lounging on the edge of fountains, huddled round small clusters of night lights, chatting."[27] To the considerable extent that public space in the United States is intended for commercial activity, it suddenly becomes useless. People caught in a blackout soon realize this and improvise new activities.

Likewise, during a blackout privately owned commercial spaces are transformed. Inside stores, the cash registers no longer work, which makes purchasing difficult. Customers chat, and conversations evolve far beyond pleasantries as the blackout continues. Soon people cease to be customers, as they realize that for some indeterminate time they are going to share the same fate. Similarly, on the streets, the pace slows, and many people loiter. In hotel lobbies, especially if the bar is well stocked, people party, and much later sleep. Dislodged from their routines, people realize that a blackout provides new possibilities. Social roles become more fluid. A businessman may take off his coat and go out into the street to direct traffic. People unable to leave a subway car or elevator break the usual code of anonymity and speak to one another. In a bar or a restaurant, the patrons continue to drink and to eat, but the pace slows, and openness to strangers increases. As long as the power is off, there is no rush, social inhibitions weaken, and hierarchies break down.

Such moments suggestively recall Victor Turner's concept of liminal space, which he developed from Arnold van Gennep's work on "rites of passage." In such episodes, detachment from a fixed position in the social structure leads into a liminal period, before returning to a stable state again at the end of the process. As so conceived, liminal space usually emerges not by accident but by design. The liminal moment is planned for, often arising at a particular time each year. Turner noted in *The Ritual Process* that "liminality is frequently likened to death, to being in the womb, to invisibility, to darkness."[28] The darkness of a blackout combined with the suspension of most normal activities suggests an experience that shares many of these characteristics. And yet, the blackout is not planned for but is only a latent possibility. It arises as a sudden violation of the "normal." Its unframed liminality occurs in the ordinary spaces of everyday life rather than within a designated ritual space. Furthermore, the desired outcome of a blackout is not a transformative rite of passage but a return to where one was before. Nevertheless, the uncertain transition period of the blackout shares some characteristics with the liminal state. Turner recognized that liminality was by no means confined to ritual moments. He wrote that the "social dramas" of a crisis have "liminal characteristics, since it is a threshold between more or less stable phases of the social process."[29] In each case, distinctions of rank disappear and individuals become temporarily homogeneous parts of a *communitas*. In contrast to ordinary life, in a liminal moment they disregard social status and personal appearance, and they become aware of the common human bond that is necessary for society to function.[30]

The disruption of a blackout contains elements of liminality, but it teeters on the edge of disaster. It is a crack in the flow of time and in the structure of space, which at best can lead to a restoration. When a power failure breaches ordinary experience, creating an immediate crisis, utilities and public services take action to sustain crucial services, repair the technical system, and return to "normal." But in contrast to the frenetic activity of utility workers and police, during a blackout most people enter a state of suspended animation, which sharpens their perceptions of their immediate surroundings and their accidental companions. The city is quieter and sounds unfamiliar. Hot, dark buildings are often less comfortable than the front stoop or the street. Neighbors who have only vaguely nodded to one another for years come out and fall into conversation.

By 1965, New York's electrical system had been reliable for a generation. Consolidated Edison had absorbed more than a dozen smaller companies between 1936 and 1960, notably Westchester Lighting Company and Yonkers Electric Light & Power in 1951. By the early 1960s a web of interconnecting lines linked the New York system with electrical generating facilities 500 miles or more away, including Toronto and the hydroelectric plants at Niagara Falls. Lightning might disrupt one part of the system, but the utilities had installed extra capacity that could be activated during an emergency, and they could use regional interconnections to shunt power quickly to where it was needed.

On November 9, 1965, Consolidated Edison was producing most of the 4,770 megawatts it required, importing only 220 MW from the north. Strictly speaking, it did not need any power from outside, as it had 1,350 MW in "spinning reserves." But the power from Niagara was an inexpensive way to meet the peak demand of late afternoon and early evening. When this power transmission was suddenly interrupted, it triggered a blackout that hit most of New York State at 5:27 p.m. The cause was a single improperly maintained circuit breaker on a Canadian high-tension line that was carrying power from Niagara Falls into Canada. The circuit breaker incorrectly responded to an increase in the load, acting as though the line was dangerously overloaded, although it was not. When it "tripped out," the energy on that line instantaneously shifted to four other lines, which then really were overloaded, and so they also tripped out.

Suddenly, a huge flow of power from Niagara Falls— 1,800 MW—could not go north. Shifting south, it overwhelmed the transmission lines that served a string of cities across upstate

New York—Buffalo, Rochester, Syracuse, Utica, Schenectady, Albany—then rushed down the Hudson River to New York City. The surge of power caused an automatic cascade that raced through these cities almost at the speed of light, shutting down all parts of the system in less than 4 seconds.[31] The disruption also cascaded into New England, darkening most of its cities and towns. It did not affect a few smaller communities that had municipal (public) power systems that were not plugged into the private grid. Hartford also escaped, because an alert operator managed to disconnect its local utility in time. New York's Consolidated Edison was only one of the many customers for Niagara power that evening, and it could have done without it by using its spinning reserves. But when the cascade cut off power in the north, it was called on to make up the whole shortfall, instantaneously. Con Ed was somehow to supply not just an additional 220 MW for its own needs, but all of upstate New York's power shortfall too. This was impossible. The very interconnections intended to ensure every local system's supply instead ensured that all of them failed simultaneously.[32]

Toronto's system also went down, but with the power available from nearby Niagara it could get its system up and running again in less than two hours. In contrast, the blackout lasted as long as 13 hours in some parts of New York City.[33] Consolidated Edison was not directly responsible, but its customers blamed it anyway, with some reason (of which more in a moment). Only Staten Island and parts of Brooklyn escaped being blacked out, because they were served from New Jersey. All television stations went off the air. A man buying chocolates watched an escalator descending, and as the lights dimmed it glided more and more slowly until the passengers came to a stop.[34] The city similarly seemed to float into stasis. Thousands of people were trapped in elevators, and a few endured long hours there. Not everything shut down. The telephone system worked on its own emergency power, some radio stations stayed on the air, and many hospitals carried on, using emergency diesel generators. Yet more than 60 of the New York area's 150 hospitals lacked adequate backup power. Though no patients died as a result of the blackout, these facilities, like most of the city, had to improvise.[35] Throughout the Northeast, doctors worked by flashlights and candles, often helping people who had fallen in the dark. (See figure 3.1.)

There were twinges of fear as people realized that the entire Northeast was dark. *The New Yorker*'s "Talk of the Town" column

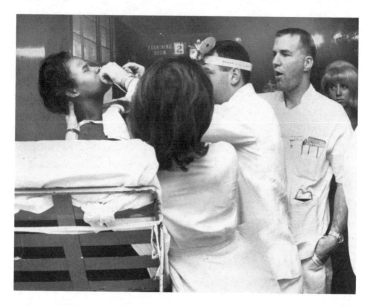

Figure 3.1
Doctors assist a young woman injured in a fall inside her darkened house in Boston. Boston Public Library.

asked: "Was there anyone whose mind was not touched, at least fleetingly, by the conviction that this was it—that the missiles were on their way, and Doomsday was at hand?" It seemed possible that "the sudden evening blindfolding, the inexplicable snatching away of light, coherence, and continuity, so closely matched the opening scene of the perpetually recurrent bad dream" of imminent disaster.[36] "Initially some people thought the blackout was to do with Soviet sabotage, or even UFOs, but there was a general sense that the government had things in hand."[37] Radio broadcasts reassured the public that the blackout did not have an extraordinary cause, and that power would be restored by the following day. A survey conducted later found that 70 percent of the population initially saw the blackout as an ordinary event, and that within half an hour 35 percent had been reassured by a radio broadcast. The blackout did not produce widespread panic or fear, primarily because people understood it as an ordinary interruption of service and because radio news helped them to interpret the event as it unfolded.[38]

Though people did not panic, some were greatly inconvenienced. Without signaling systems, railroads were unable to operate. (The Pennsylvania Railroad, which was on the New Jersey grid, kept running.) Tens of thousands of shoppers were caught inside stores; some remained there for hours, or even all night. People slept in hotel lobbies, in ballrooms, and in public spaces. An estimated 60,000–80,000 people were trapped in subway trains. Evacuation took hours, and at midnight 10,000 were still stranded underground. Before the subway system could be started again, every line had to be patrolled and inspected. One journalist recalled "a piece of television news which showed some people who'd been stuck all night in a subway car. A crowded mixed bag of young and old, well dressed

Figure 3.2
Commuters stranded by the power failure settle down for a long wait on the steps of New York's Commodore Hotel, November 9, 1965. Associated Press.

and shabby, they seemed absolutely overjoyed at their predicament. And when the subway policeman who had been stuck with them congratulated them on the way they had been good comrades and otherwise passed the time courageously, they cheered him wildly."[39]

The public generally behaved well. "Although persons in public vehicles (other than trains) and at work tended to regard the blackout as somewhat more unusual than others, there is no evidence that they perceived any real or potential threat in the situation."[40] Travelers at airports found themselves in unlighted buildings looking at dark runways from which planes could not land or take off. At LaGuardia Airport, one runway was cleverly re-lighted with auxiliary power from a water pump. Although 240 aircraft landed successfully, more than 250 other flights were cancelled or diverted.[41] For people flying at the time of the blackout, the experience was surreal. Passengers invited to gaze at the lights of Manhattan saw them flicker out. It clearly was not fog or bad weather that obstructed the view—the lights of automobiles remained visible, tracing the lines of the streets. But the skyscrapers had become indistinct hulks, and the famous electric signs were gone. Rush-hour drivers found it hard to get home. Traffic signals stopped working, and drivers who made their way to the Queens Midtown and Brooklyn Battery tunnels found them closed, for they lacked lights to guide traffic and fans to evacuate the exhaust. (The Lincoln Tunnel, the Holland Tunnel, and the George Washington Bridge got their electricity from New Jersey.) Because ships had autonomous electrical systems, the harbor was far brighter than the rest of the city (figure 3.3).

A few people carried on regardless of the lack of power— notably the pianist Vladimir Horowitz, who continued a rehearsal, first playing from memory and later with the aid of flashlights. Many made extraordinary efforts. Zookeepers found ways to keep their animals warm on a cold November night. In general, public order was good. At Walpole Prison in Massachusetts, 300 inmates rioted for two hours under the cover of darkness, and in a few cities there were scattered reports of petty crime or vandalism. But the 12,000 policemen on duty in New

Figure 3.3
During the 1965 blackout, a freighter shines in Brooklyn but Manhattan is shrouded in darkness. Library of Congress.

York found that the crime rate was lower than normal. There were four times as many fires as usual, however, because people tried to improvise heat and light.

After initial fears subsided and people understood the situation, many people experienced unexpected elation. Staffers at *The New Yorker* noticed the strange beauty of the city: "The moonlight drew us back to the window, and as we stared out at the impossible, unimaginable loveliness of the moonstruck towers rising out of the blackness, and now just beginning to flicker with tiny bug glimmerings from fellow-survivors, we were abruptly and unaccountably elated."[42] They realized they were caught not in a catastrophe but in an adventure. Though the blackout persisted all night, the public was almost uniformly peaceful, relaxed, and even exuberant. A party mood prevailed, as the news media later described in detail. A study of the public, commissioned by the Office of Civil Defense, found that fear was not a widespread reaction and that it did not prove contagious. The investigators were surprised, because "projection of one's own fear on others is a fairly well-known phenomenon which has been experimentally induced." Instead, they found a "contagion of joy during the blackout" that "does not appear to have been accompanied by a corresponding contagion of fear."[43]

The weekly magazine *Life* devoted a whole issue to the experience, concluding that "the blackout quite literally transformed the people of New York," taking away their anonymity. The columnist Loudon Wainwright Jr. described the unexpected public intimacy and the party mood:

Ordinarily smug and comfortable in the high hives of the city where they live and work, they are largely strangers to one another when the lights are on. In the darkness they emerged, not as shadows, but far warmer and more substantial than usual. Stripped of the anonymity that goes with full illumination, they became humans conscious of and concerned about the other humans around them. In the crowded streets, businessmen, coats removed so that their light-colored shirts could be seen, became volunteer cops and directed traffic. Though the sidewalks were jammed, there was little of the rude jostling that is a part of normal, midday walking in New York. In the theatrically silver light of a perfect full moon (a must for all future power failures) people peered into the faces of passersby like children at a Halloween party trying to guess which friends hide behind which masks. In fact, the darkness made everyone more childlike. There was much laughter, and as they came down the stairs of the great office buildings in little night processions led by men with flashlights and candles, people held hands with those they could not see.[44]

Networks of urban conveniences usually eliminate the need for much personal contact or fellowship. But the 1965 blackout forced New Yorkers back into touch, demonstrating their interdependence and temporarily making them gregarious. Numerous accounts of the blackout provide similar stories of the public reaction, which accords well with the idea of a liminal event in which hierarchies break down and people from all levels of society feel united. Many people responded in this way during subsequent blackouts, and those who have experienced a blackout generally regard it as a remarkable experience that was not necessarily unpleasant despite some hardships. "Since liminal time is not controlled by the clock," Turner observed, "it is a time of enchantment when anything might happen."[45]

The public discovered that a blackout could abet an unexpected moment of spontaneous cohesion. In a showroom for glass and crystal, customers stood still in the darkness, taking care not to break any of the expensive merchandise. After employees lighted candles and reestablished a sense of orientation, "customers were surprised when the sales force began bringing out an elaborate dinner and arranging it on long glass counters. There was chow mein, turkey, fresh fruit, shrimp and sweetbreads—all prepared in advance for a sales meeting that became, instead, the Blackout Ball."[46] At Forty-Fifth Street and Sixth Avenue, lines of people waiting for buses were entertained by two boys playing bongo drums. On a stalled subway car, no one talked for about 20 minutes; then a conductor said that people could smoke if they wanted to, even if they were nonsmokers. This provoked a laugh, and "soon everyone was talking and getting acquainted."[47] A mock religious procession marched along Sixth Avenue singing "Hark the Herald Angels Sing."[48] People trapped in elevators "sought relief in community singing.'"[49] A woman returned home to her apartment near Union Square and found all the neighbors gathering in the one flat that had a gas stove. Neighbors who had seldom spoken brought out choice items from their refrigerators for an impromptu party, which established friendships that persisted for decades.[50] At one department store, management was able to find two charter buses to take employees home. "Someone thought it would be fun if they danced out of the store," so out "came a human chain, dancing."[51] People also jauntily made the most of the moment in Boston and other major cities. (See figure 3.4.)

Figure 3.4
In Boston's Sheraton Hotel, the Café Riviera serves dinner by candlelight. The food was cooked on gas stoves. Boston Public Library.

The lighthearted reactions to the 1965 blackout were also suggested in the MGM film *Where Were You When the Lights Went Out?* Its poster referred to "the liberties that were taken the night New York flipped its fuse and became 'Fun City'" and the "8 million New Yorkers who were lost in the dark . . . until they found each other."[52] On the poster, a scantily clad Doris Day parodied the Statue of Liberty, holding a large candle instead of a torch. Her other hand clutched a book titled *The Constant Virgin*. Behind her, New York's skyscrapers were dark. Couples in intimate conversation stood in the foreground. The rather thin plot builds up and resolves a series of comic misunderstandings and leads to lovemaking during the blackout. Nine months later, the Doris Day character gives birth.

An urban legend grew that lovemaking had been rampant during the blackout and that the birth rate had spiked 9 months later. There seemed to be some logic to that. Unable to watch television or read, and perhaps a bit frightened, surely many couples became intimate in the gloom. They might not have been able to find contraceptives in the dark. Perhaps some people trapped with attractive strangers enjoyed one-night stands. On August 10, 1966, the *New York Times* ran a page-one story headlined "Births up 9 months after the blackout," but subsequent demographic analysis covering the years 1961–1966 found a normal birth rate for the day and the week in question.[53] The persistence of this story in popular memory testifies not to an increase in copulation, but to the sudden intimacy with strangers and the liminality of an event that in many ways felt like a gigantic party.

It is instructive to compare the 1965 blackout with another famous moment of social solidarity in the history of New York City: the rededication of the Statue of Liberty in 1986.[54] For that carefully choreographed civic event, the organizers provided spectacular laser lighting and an immense fireworks display (40,000 projectiles, 100,000 bursts) that was seen by several million people who lined the harbor and filled much of it with small boats. The many security guards and extra police on duty, prepared for unruliness or trouble, had almost nothing to do. As the *Daily News* put it in a headline, "Nightmare just never happened." The *New York Times* noted the crowds were "notably serene, almost reverent, as if paying homage to something righteous and inviolable." One participant told a reporter: "There's a brotherhood and everybody's friendly." In this case, citizens expected a rupture in ordinary experience. Public officials organized a sublime event culminating in the spectacular relighting of the torch of the Statue of Liberty, which induced a sense of unity and patriotism. At the climax of the display, "strangers embraced one another."[55] In this liminal moment, as in the 1965 blackout, social divisions broke down. But the 1986 ceremony was carefully structured in advance. Rededicating the Statue of Liberty had affinities with the planned rituals that Turner analyzed. In contrast, a blackout was by definition unscheduled and unscripted. The crowds at the Statue of Liberty in 1986 were excited and amazed not by the accidental loss of illumination, but by a super-abundance of dazzling special effects. The sheer surplus of energy became both a visible sign of patriotism and a stimulus to revelry.

Why was the public response to the 1965 power failure so affirmative? To answer this question, it is useful to look beyond Turner's work to another theory of unscripted breaks in social time. Blackout experiences can also be understood as examples of what Michel Foucault calls "heterotopia." Foucault specifies that "heterotopias are most often linked to slices in time," and that they begin "to function at full capacity when men arrive at a sort of absolute break with their traditional time."[56] Paradoxically, the "traditional" time of the modern city is based not on natural rhythms but on acceleration and the pressure to compress more experience, more work, and more "reality" into every hour of the day. The blackout forcibly breaks this pressure, stopping almost all public clocks and preventing most work activities. It interrupts the flow of both production and consumption, and confronts people with timelessness, with living and improvising in the now. The blackout breaks with modern, capitalist, productive time, and with the simultaneity of electrical systems. If accelerated time is money, then the timelessness of a blackout uncovers non-monetary values and other uses of space.

Foucault argues that every heterotopia is "capable of juxtaposing in a single real place several spaces, several sites that are in themselves incompatible."[57] In a blackout, the electrified city is juxtaposed with a darkened twin of itself. The city stripped of its power system has the same contours and occupies the same physical location as the electric city, but it is a different place, and it both looks and sounds different. The continual drone of the city—a composite white noise from millions of air conditioners, blaring loudspeakers, rumbling subway cars, and electric machines of all kinds—suddenly stops. Simultaneously, the illuminated city of skyscrapers and spectacular electric signs disappears, revealing a landscape that inverts the values embedded in the electrified city. The most obvious of these values is the belief in progress, which electricity continually embodied at world's fairs, notably those held in New York in 1939 and 1964. At the World's Columbian Exposition (Chicago, 1893), the electrical building was one of the most popular sights. Electricity served as the central theme of the Pan-American Exposition

(Buffalo, 1901), which celebrated the opening of hydroelectric plants at Niagara Falls.[58] For the 1939 "World of Tomorrow" Exposition, in Queens, Consolidated Edison erected an exhibit building a city block long that contained a giant diorama of New York City as it might look as an electrified utopia in 1960. In a portrayal of the "Forward March of America," visitors saw a recreation of a cobblestone street of 1892, with dimly lighted stores, an ice wagon, a horse-drawn streetcar, and gas street-lights. They were prompted to think of the electrical improvements to domestic work by glimpsing housewives laboriously doing their laundry by hand or cooling themselves with hand-held fans. From this "Street of Yesterday" visitors passed to a brilliantly illuminated "Avenue of Tomorrow" with smooth asphalt streets, sleek automobiles, department stores, and sky-scrapers. Such exhibits encouraged visitors to equate national progress with the spread of electrification. In case they missed the point, a quotation from Thomas Edison displayed on the wall near the exit proclaimed "Great days are ahead and electric-ity will have a great part to play, granted only that it can be unfettered, with full opportunity for the largest possible indi-vidual initiative and energy."[59] In the first 100 days more than 6 million people passed through the Consolidated Edison pavil-ion, and First Lady Eleanor Roosevelt praised it in a newspaper column.[60] After World War II, the same themes continued. During the 1950s and the 1960s, General Electric continually ran advertisements declaring "Progress is our most important product," and Ronald Reagan regularly delivered this slogan on television as the host of the General Electric Theater.[61] For nearly a century, the electrified city was presented as the prime example of a dynamic utopian landscape. The un-electrified city was its dark, static, regressive twin.

The electrified city in general and New York in particular was intimately associated with the modern in art and architecture. Through the paintings of Joseph Stella and Georgia O'Keefe and the photographs of Edward Steichen and Alfred Stieglitz. Americans had learned to see New York's almost cubist skyline as the visualization of modernity.[62] For decades, postcards had celebrated the scintillating night city, to the point that it seemed natural and normal. Not only the skyscrapers but also the city's bridges, public buildings, and monuments were bathed in light. A blackout darkened the skyline, transforming it into a hulking, static cliff. New Yorkers were suddenly disoriented as landmarks disappeared or took on an entirely new appearance.

When a blackout defamiliarizes daily surroundings, many people get out their cameras, wanting to capture the novel scene. Like tourists, they usually resort to conventions of representation, as familiar photographic genres come to mind. In the nineteenth century the dimly lighted night world was a common subject in the "nocturne"—a genre favored, for example, by James McNeill Whistler, notably in "Nocturne in Black and Gold, the Falling Rocket" (1875, Tate Gallery). Such images directly influenced pictorialist photography, which became prominent in the 1890s, during precisely the years in which cities were being electrified. Alfred Stieglitz, Alvin Langdon Colburn, and other photographers depicted the city at night as a pattern created by the interplay of natural and electric lights. At times, as in celebrated images of the Flatiron Building or the Singer Tower, the soft night glow of the city provides the background for an impressive man-made structure. Such photographs celebrate both the skyscrapers and the networked systems that sustain them. Often, these photographs look down from a tower, so the city appears as a vast, coherent landscape.

Some photography of electrical blackouts resembles this older pictorial tradition, but more commonly they recall the late-modernist photography of Alfred Stieglitz. Though Stieglitz's 1932 image reproduced here as figure 3.5 was not of a blackout, it seems surprisingly dark to twenty-first-century eyes—a reminder that an artistic image taken with quite other intentions may come to have an unintended documentary value. Such dark buildings became a common blackout motif, particularly seen from a distance, when the city skyline seemed either a dark escarpment or a series of lifeless towers. The cover of an issue of *Life* published after the 1965 power failure typified this approach, which remained common in subsequent blackouts. (See, e.g., figure 6.2 below.) Such visual imagery reinforced the sense of blackouts as extraordinary breaks in urban space and time.

These images also emphasize another value that disappears during a blackout, a value that is deeply embedded in the electrified city: incessance, or the possibility of continuous activity, based on ubiquitous energy. Darkness had been eliminated as an impediment for those who wanted to work, shop, stroll, or exercise at night. Electric current could be brought to any location, flexibly and easily, whether underground, atop a skyscraper, or to an airplane miles overhead. With electricity, it seemed, anything could be done anywhere, at any time. Americans always had radio and television on the air. Some stores were always open, and many activities were available around the clock. If hyperactive, brightly lighted cities were novel in the early twentieth century, they seemed a necessity two generations later. Loss of power was no longer merely inconvenient. People confronted with a blackout lost an enormous range of possible activities. A blackout was not merely a technical failure;

Figure 3.5
Alfred Stieglitz, "From My Window at An American Place Southwest,"
1932. Board of Trustess, National Gallery of Art, Washington.

it was a breakdown in the flow of action, in the city's continuity. To borrow a term from literary criticism, it was an *aporia*, a gap, a discontinuity. Through this crack in social time and space bubbled up new experience.

The meaning of blackouts changed markedly during the twentieth century. In 1936, the electrical system had not yet been entirely assimilated into everyday life. It was not yet "natural." Indeed, the military manipulated its artificiality to create the appearance of darkness. By 1965, however, many New Yorkers regarded a blackout as a violation of the expected order of things. It seemed an anomaly, but one with no long-term implications. It was dislocating, but not upsetting. As a survey found afterwards, "there is no evidence that they perceived any real or potential threat in the situation."[63] Because the blackout was not thought to be dangerous, the paralysis of such a night could become a liminal moment. In later years, however, both the perception of and response to blackouts would shift. As they lived through a succession of power failures, urban Americans gradually understood how vulnerable a city could be if its energy system were to break down. They realized that in a week, or even less, urban life without electricity could become unsustainable. This realization was not central to the experience of the blackouts of 1936 and 1965, but emerged more clearly as part of their meaning in 1977 and 2003. These later events would also demonstrate that the unexpected breakdown of the electrical system could prompt behavior quite unlike that in the 1965 blackout. Rather than bringing people together, a major power loss could aggravate tensions in the community, lead to unrest, or unleash criminality. The public might also express depression, fear, or anger at public officials or the power company. However, such behavior was not anticipated after the experience of 1965. Electricity had been woven into every aspect of everyday life, yet it seemed the community could pull together and do without it in a pinch. A blackout was perceived not as a social or cultural crisis but as a temporary technical problem.

In the aftermath of the 1965 blackout, observers found it "incredible that there was no adequate auxiliary electricity available to maintain at least minimum essential services at Kennedy International Airport, at the Empire State Building, the Pan Am Building and other major skyscrapers, or even at some hospitals, or on the subway and commuter railway."[64] All these institutions were upgraded. Meanwhile, utility executives and government officials took steps to prevent a similar event. The utilities voluntarily formed regional reliability councils, which were represented on a North American Electric Reliability Corporation (NERC). These organizations were to ensure that the electrical system would be better designed and coordinated. They exchanged data and conducted simulations to determine how the grid would behave in unusual situations and emergencies. NERC was created both to avoid the high cost of blackouts and as a corporate defense against government involvement in the largely private power industry. If utilities had not established something like it, the federal government might have stepped in and supervised them.

Government and private enterprise agreed that the problem was not too much complexity or too much integration, but too little. Each individual utility was urged to consider ways to make its own infrastructure more robust. In response to increased public scrutiny and criticism, utilities spent more money on training their operators, and installed more load-shedding

devices that could respond to a sudden fall in supply by automatically cutting off some customers, temporarily, to balance
supply and demand. As utilities added these new layers of control and monitoring, they tied the grid more tightly together.
After taking these measures, the industry hoped that nothing
like the great blackout of 1965 could ever recur. Experts on risk,
however, have found that "normal accidents" cannot be anticipated.[65] The more complex a system becomes, the more small
errors and inconsistencies accumulate and incubate until they
hatch as unforeseeable problems. If NERC could anticipate
many specific blackout scenarios, it was difficult to anticipate
all of the possibilities and to coordinate thousands of technicians and power plant operators, each seeking to control just
one part of a system that covered thousands of square miles.
Not only did they have to shunt power instantaneously from
place to place as local demand rose and fell; they also had to
respond to unexpected shocks to their system—ice storms, lightning strikes, tornados, hurricanes, fires, and other disruptions.
Through all the daily fluctuations in demand, as they brought
generators on and off line, all parts of the system had to be kept
synchronized at precisely 60 Hertz. The utilities performed this
feat 24 hours a day for years on end, with few interruptions, but
they could not anticipate every contingency. Any local "normal
accident" conceivably could begin a cascade. Ominously, after
the 1965 blackout, the vice-president in charge of engineering
at Boston Edison declared in a newspaper interview that if there
had been a national grid, the whole country might have been
"plunged into darkness in less than a second."[66]

In July 1977, not quite 12 years after the 1965 blackout, New York City again had a power failure, but the public response was far different. The earlier blackout had occurred on a mild November night, and it had been pleasant to be out in the streets. But during the 1977 blackout the temperature rose into the nineties, and parts of the city erupted in arson, looting, and riots. The sense of community had broken down, and the blackout hardly induced a liminal moment of unity. It revealed a fractured society. Not only had the national prosperity of 1965 evaporated, but New York City had suffered a major economic reversal. More than 600,000 jobs had left the city between 1969 and 1977.[1] The city teetered on the edge of bankruptcy, and an Emergency Financial Control Board dictated its budget. It imposed severe cutbacks on public services, closed hospitals and libraries, discharged 3,400 police and 1,000 firemen, and froze the wages of city employees.[2] Crime rose dramatically, arson "went unchecked," and "an understaffed [fire department] barely kept up. In 1975, there were 13,000 fires reported in the Bronx alone."[3] (See figure 4.1.) In short, the 1977 blackout occurred during a hot spell in a poorly policed and nearly

Figure 4.1
Firemen battle flames in the Bronx, July 14, 1977. Associated Press.

bankrupt city that suffered from high unemployment, inflation, and a general sense of social crisis.

Furthermore, in the early 1970s prices for all forms of energy had begun to rise, and there were shortages of gasoline and heating oil. Relative to other industries, electrical utilities are particularly sensitive to fuel prices, which represent "approximately 50 percent of the annualized cost of electricity produced and delivered."[4] As late as 1977 the United States burned 631 million barrels of oil a year to produce 17 percent of its electricity, and New York's system relied more than most on oil-burning plants. The price of crude oil had more than quadrupled. Drivers who could remember paying 35 cents a gallon for gasoline in the 1950s saw the price soar to more than $2 a gallon. Utilities shifted to coal to get a reliable, domestic energy supply, but they still faced rising costs, as the price of coal shot up

500 percent or more between 1970 and 1978.[5] Yet coal was still cheaper than oil and more easily obtained. The popular culture of the period was rife with conspiracy theories, often attributing high energy prices to skullduggery by the oil companies in combination with the more overt machinations of the Organization of Petroleum Exporting Countries.[6] New York City in particular suffered from high energy prices, and Consolidated Edison had the highest electrical rates in the US.[7] The 1965 blackout had been perceived as an aberration in a prosperous economy with an exemplary infrastructure. The one in 1977 was perceived as part of an ominous energy crisis that confirmed that social, political, and economic structures were in disarray. Energy shortages were not responsible for the blackout, but they provided the cultural context within which it was understood.

The energy crisis of the 1970s is often recalled as a problem of supply. Abundant fuels and declining electricity prices had been the norm for two generations, but in the early 1970s suddenly it seemed there was not enough oil or natural gas. Americans had difficulty understanding that their high-energy way of life was partly responsible. Instead, they blamed government, utilities, international oil companies, or Middle Eastern oil producers. However, the shortages were, to a considerable degree, due to demand. Residential use of electricity tripled between 1942 and 1952, then increased 250 percent in the next ten years. By 1962 American homeowners used 226 million kilowatt-hours a year, and they doubled this consumption again in just eight more years.[8] Between 1973 and 1977, when the energy crisis reduced this growth to 3.3 percent a year (annualized), this sluggish increase felt like a shortage to a people who had at least doubled their consumption of electricity every decade since 1910.

A sense of perpetual abundance had been built into the habits of consumption. As early as 1969, the magazine *Fortune* noted the dangers of brownouts and utility overloads and commented that "the American likes his home brilliantly lit, of course, and he has a passion for gadgets that freeze, defrost, mix, blend, toast, roast, iron, sew, wash, dry, open his garage door, trim his hedge, entertain him with sounds and pictures, heat his house in winter, and—above all—cool it in summer."[9] The idea that extensive energy use could be problematic did not become widely apparent until a few years later. "Americans," Secretary of the Interior Rogers Morton had noted at a meeting of the Organization for Economic Cooperation and Development, "have long been accustomed to abundance. Scarcity of natural resources and scarcity of land have not been factors to contend with. That we no longer have the luxury of unbounded clean land, air, and water, nor of the fuels that we blindly depend upon to give us pleasures in life, is a concept difficult for our general public to grasp."[10] Yet the high-energy American way of life created the rising demand that underlay the energy crisis.[11]

As early as September 23, 1970, several years before the oil embargo by the OPEC countries, the Northeast experienced brownouts during an early fall heat wave. The following year, President Nixon warned Congress that the United States was entering a period of increasing energy demands and short supplies. Politicians focused primarily on the rising cost of oil, but all energy markets are interconnected. When oil prices rise, price-sensitive consumers substitute another fuel. Some utilities replaced oil-fired power plants with coal-fired or atomic plants in the 1970s. Some homeowners switched from electricity to natural gas for heating and cooking. But overall, the United States did not reduce its energy consumption. All forms of energy use increased during the 1970s, and electricity consumption increased by 50 percent as Americans adopted electrical appliances and gadgets of all kinds. In the 1950s the typical American family acquired its first television. In the 1970s, most families had two televisions, some three. Electric clothes dryers and stereo systems also could be found in most middle-class homes.

The greatest factor in increased electrical consumption, however, was air conditioning. which began to place huge demands on the electrical grid in the 1970s. Air conditioning emerged before World War I, when a few companies started selling devices that could regulate both a room's humidity and its temperature.[12] These were particularly attractive to textile mills, where threads snapped if too hot and dry.[13] The technology spread to other sites where both temperature and humidity were crucial to production, such as breweries and factories that made pasta, candy, and cigarettes. In the 1920s the general public began to enjoy this new technology in stores, theaters, restaurants, and trains. By the 1930s, General Electric and many utilities anticipated enormous potential sales of air conditioning in the domestic market, but it remained too expensive for most consumers. Some early air-conditioning firms sold shoddy machines that did not regulate humidity but only cooled. As a result, the public perception of air conditioning remained confused. In 1938, less than one home in 400 had installed it in even one room.

After World War II, however, air conditioning became common in new offices and apartments, many of which would have been quite uncomfortable without it. In trend-setting modernist buildings that had glass exterior walls, floor-to-ceiling windows increased the cost of air conditioning by

50 percent. As sales increased, the price of units came down. Manufacturers worked with architects to design homes for their equipment. Builders found that they could partially offset the cost of installing air conditioning by eliminating attic fans, window screens, and high ceilings. Without them, however, a house was comfortable in the summer only when the air conditioning was on—particularly in the South, where generations of clever architectural adaptation to the hot climate were rapidly abandoned. These included the "dog-trot" cabin with its central breezeway, the porch that wrapped around a house, strategically planted groves of trees, floors raised a yard or more above the ground, and much more.[14]

The federal government stimulated the shift to air conditioning. In 1955 it adopted a temperature policy that required air conditioning in most of its new buildings. Also, the Federal Housing Authority authorized mortgages on homes that included it. At first, window-mounted units were the norm, but central systems gradually became standard. By the 1960s, they predominated in Las Vegas, Dallas, Phoenix, Bakersfield, and other Southwestern cities. Nationally, domestic installations went from 6.4 million in 1960 to 24 million in 1970.

In the 1970s, air conditioning rapidly shifted from being considered a luxury to being considered a necessity. It created an enormous summer demand for electricity. New homes often were designed to minimize purchase price, not operating costs. Sealed windows trapped heat inside the house even as hair driers, toasters, stoves, televisions, radios, refrigerators, and electric lights produced more heat. Such houses made air conditioning a necessity, and the resulting power demands exacerbated energy shortages. By 1982, immediately after the "energy crisis," over 60 percent of American homes had air conditioning, and half of these had central systems for cooling the entire house. Insofar as this technology made little sense in the Pacific Northwest, in the North, at the higher elevations, or in many coastal communities, this was high market penetration.[15]

As Los Angeles, Houston, Phoenix, and other high-energy cities sprawled, they adopted glass-walled office towers that were unendurably hot without air conditioning. Fluorescent lighting, perfected at the end of the 1930s, reduced heat buildup somewhat, but new office equipment, such as computers, photocopiers, and fax machines, added more heat. The characteristic (if often characterless) modern office building, with its sealed, tinted windows, was habitable only if it had electricity. The soundscape of such buildings included the constant swoosh of cool air, whirring elevators, humming computers, and buzzing fluorescent lights. These modernist buildings also had ceilings lower than those in older skyscrapers, and they packed nine stories into the same height once used for seven or eight. Less adaptable to the climate than most greenhouses, which typically open their windows automatically to combat overheating, glass office towers and apartments overheated quickly and had to be evacuated during any summer power failure.

Air conditioning fundamentally changed the demand for electricity, which from 1880 until 1960 had peaked during the winter and during the darker periods of the day, with lower demand on sunny days. The demand for air conditioning was highest in summer, and it peaked during the brightest hours. Initially, utilities were delighted by this new load factor, which evened out daily and seasonal demand. In the 1970s, however, air conditioning began to strain the generation and transmission system. By the 1990s, both the utilities and the public knew that the most likely time for a blackout was no longer in the dark of winter but during a summer heat wave. In that sense, the New York blackout of 1977 was a transitional event, coming in the summer, but at night, and caused not by excessive demand, but by lightning.

As the example of air conditioning illustrates, by the end of the 1970s, in a thousand ways, the high-energy consumer was deeply embedded in electrified structures. Most obvious was the detached suburban home filled with ever more appliances and connected to an energy-intensive system of services, including telephone, radio, television, and (incipiently) the computer. Just as important were air-conditioned shopping malls and domed, climate-controlled sports stadiums with artificial lighting, enormous electronic scoreboards, and instant-replay screens. Americans not only took such energy-intensive sites for granted, they expected them to improve and expand.

Intensifying energy use also transformed urban weather. During the nineteenth century, scientists had begun to detect small temperature differences between cities and their immediate hinterland.[16] By the 1950s, each major city created a noticeable "heat island." Asphalt parking lots and roads soaked up sunlight, glass buildings trapped heat, and both kept radiating warmth after sundown. More heat came from hundreds of thousands cars and trucks, especially in automotive cities such as Houston and Los Angeles. Still more heat came from factories, power plants, and the exhaust from air-conditioning systems. In Houston, the accumulated result was a daily temperature 5–9 degrees (Fahrenheit) higher than that of the surrounding countryside.[17] In some cities the differential could be as much as 10 degrees, and this differential extended into the evening hours. The hotter it got, the more people wanted air-conditioned cars, offices, shopping malls, and homes. Indeed, Houston built the first air-conditioned sports stadium, the Astrodome. Thus Americans used electrification in general, and air conditioning in particular, to create a self-reinforcing pattern, or feedback loop. The more energy they used, the higher urban temperatures soared, creating demand for even more energy.

In each city the heat island was a local effect, but increased greenhouse gases (carbon dioxide, methane, etc.) also began to raise global temperatures. While an average increase of one or two degrees might not be that noticeable, global warming was not expressed as a slight average rise. Rather, the weather became unstable, with greater swings in temperature, including extremely hot spells. Each torrid night increased the attraction of owning an air-conditioned house, and the perceived need for it intensified as consumers grew more aware of global warming.

Not all citizens had equal access to air conditioning. Poor neighborhoods and slums quite literally got hotter as a direct consequence of the white middle-class flight to the air-conditioned suburbs, which demanded more heat-absorbing roads, larger parking lots, and higher overall energy use. In Los Angeles, as irrigated orchards were replaced by houses, roads, and suburban sprawl, summer temperatures rose dramatically. The demand for electricity increased even faster: consumers wanted 2 percent more power for every increase of a single degree (Fahrenheit). At the same time, the amount of smog in the air rose 3 percent for every degree above 70.[18] Greater use of energy by the middle class created a feedback loop that disproportionately affected minorities and the poor, driving up temperatures and increasing air pollution in the heat island of the central city. As the urban core became hotter and its air grew dirtier, the contrast with suburban, air-conditioned life was ever more palpable.

Furthermore, those who grow up with air conditioning lose some of their ability to cope with heat. Mark Blumberg concluded in his 2002 book *Body Heat* that children who grow up in an air-conditioned environment lose some of their capacity to sweat.[19] As adults, they cannot cope with heat waves as well as those who have been acclimatized. They are literally dependent on air conditioning. "If you stay in your chair in an air-conditioned room and never have to turn on your hormones and heat-loss mechanism," Blumberg wrote, "you're not going to be a fit organism."[20] Increasingly, suburban Americans did not make much use of their "heat-loss mechanism." Instead, they moved from one air-conditioned environment to the next.

As high electrical demand was being built into human physiology and the "necessities" of homes and offices, the ability to produce electricity cheaply was coming to an end. Before the 1970s, energy costs had remained low. Engineers had continually found more efficient ways to burn coal, and they had achieved remarkable economies of scale. They had installed larger and more efficient steam turbines that operated at higher pressures and higher temperatures. General Electric, Westinghouse, and other boiler and turbine producers adopted new metal alloys that could withstand the corrosive effects of superheated steam. Improvements in metallurgy also made possible larger turbines that could withstand higher pressures. Expanding regional grids made operating extremely large turbines a practical and economical choice. However, by about 1970 steam technology had reached technical limits. A few gigantic steam turbines were installed, the largest generating 1,400 megawatts, but plants rarely exceeded 1,000 MW, because there were no further gains in efficiency, which topped out between 40 and 45 percent.[21] Between 1920 and 1970, economies of scale permitted the price of electricity to fall from 7.5 cents to just over 2 cents a kilowatt-hour. The coal burned to produce a kilowatt-hour decreased by 70 percent between 1920 and 1959.[22] However, by 1970, as documented by Richard Hirsh, engineering ingenuity was running up against physical limits, notably "metallurgical weakness at high temperatures and pressures" and the difficulties of using water to cool very large units. Likewise, turbine blades could not simply be scaled up to ever-larger dimensions; they began to crack and under-perform beyond a length of $28\frac{1}{2}$ inches.[23] Overall, beyond certain steam pressures and beyond a certain size, there were no further efficiencies.[24] At the same time, utilities realized that building gargantuan power plants resembled putting all one's eggs in a gigantic basket. When a 1,000-MW plant went off line for maintenance, its entire generating capacity had to be found elsewhere. Once the industry had reached these technical and practical limits, the cost of electricity could fall only if the price of fuel fell. But after 1970 fuel prices escalated. Consumers paid twice as much for electricity in 1975 as they had in 1970, and the price doubled again by 1980.[25]

Americans briefly discovered the virtues of conservation. Symptomatically, the Tennessee Valley Authority "instituted programs for home insulation and industrial energy conservation" to reduce the need for additional generating facilities. Private utilities also began to use load management as a substitute for building new power plants. They did so in part because environmental groups and government regulators often challenged new construction.[26] Yet if the engineering profession and many utilities learned the long-term economic benefits of being more energy-efficient, consumers responded primarily to price. When energy costs stopped rising in the 1980s, old habits reasserted themselves. The consumer wanted a larger refrigerator and a larger TV set, and often was only vaguely aware of whether new appliances consumed more or less power.

During the 1970s, Americans paid more attention to energy issues. Doomsday scenarios circulated. Pundits predicted that future wars would be fought over energy, not ideology. Paul Ehrlich's bestseller *The Population Bomb* predicted imminent worldwide starvation and a life-and-death struggle for basic resources. Ehrlich spread his warnings through radio and television interviews as well as through his book. Another popular work, *Energy Crisis*, predicted "strip-mining for coal on a vast scale in the US" in the early 1980s, a "Great Depression of 1929 scope," and "a stock market collapse." By the end of the 1980s there would be "a world conflict over energy resources," and increased pollution. As a result of energy shortages, the US would lose "world leadership to Russia." The book also predicted "a hotter world climate" caused by increasing carbon dioxide with "massive and unpredictable environmental consequences."[27] In *The Limits to Growth*, a team of scientists and social scientists explored the complex interactions between industrialization, increasing consumption, population growth, and the environment. They concluded that unless pollution levels were reduced and growth restrained the world could be pushed into ecological collapse by 2100.[28] Jeremy Rifkin, Barry Commoner, and other environmentalist critics also believed that the whole edifice of the existing economic order was crumbling.[29]

The sense of crisis did not arise solely because of these criticisms from academics and environmentalists. In the spring of 1977, President Jimmy Carter made energy conservation the focus of his first major legislative initiative. Turning down the heat in the White House, Carter urged Americans to improve their home insulation and to buy automobiles that got better mileage. "The cornerstone of our policy," he declared, "is to reduce the demand through conservation." The belief that the United States faced long-term energy shortages was thus at the forefront of public discussion in the months immediately before the 1977 blackout.

The oil companies, which made large profits in the 1970s, found that the best way to defend themselves against charges of profiteering was to assure the public that the crisis was quite real and that it required enormous investments for exploration and difficult extraction.[30] Many advertisements emphasized the "expensive and risky gamble" involved in discovering and extracting oil overseas.[31] Exxon and Chevron stressed their efforts to extract oil from far beneath the sea, from old fields using new recovery methods, and from new sources, such as oil shale.[32] Utilities likewise explained higher prices as the result of energy shortages. For example, Southern California Edison sent a lavishly illustrated 20-page booklet titled "The Energy Crisis" to its customers. "The Energy Crisis" presented the problem as one of supply, and pushed for nuclear power as the "solution."[33] The American Electric Power System, a group of power-generating companies in Appalachia and the upper Midwest, ran full-page advertisements advocating greater use of American coal, "our ace against Middle East Oil."[34] They argued that an overzealous Environmental Protection Agency prevented use of Eastern bituminous coal, while the Department of the Interior refused to release "millions of tons of clean Western coal."[35] In short, energy producers and the federal government agreed that the energy crisis would persist.

In 1977, amid rising electricity prices, energy shortages, and doomsday scenarios, New Yorkers experienced a blackout quite unlike that of 1965. A power outage marked by looting and disorder seemed not a temporary disruption of plenty but the foreshadowing of a future of rationing.

However, the 1977 blackout was not caused by energy shortages, even if they did provide a contextual frame for the event. Lightning initiated the outage, which could have been avoided if Consolidated Edison's circuit breakers and grounding system had been properly maintained.[36] When lightning struck two 345-kilovolt power lines, each should have been out of commission for only a few seconds. Instead, both went completely out of service, triggering a cascade of other malfunctions. Most seriously, the Indian Point nuclear power plant automatically shut down because there was no outlet for its generators. Even so, at first the system coped with the emergency by bringing in additional power from elsewhere on the grid. However, less than 20 minutes later lightning struck again and short-circuited two additional 345-kilovolt lines. Again circuit breakers failed to function properly, and Consolidated Edison's largest steam turbine ("Big Allis") automatically went off line. Investigators later conceded that being struck by lightning twice in rapid succession was bad luck, but the blackout occurred because equipment malfunctions put four major transmission lines and two major generating plants out of service. A huge overload stressed the system that remained. Almost exactly an hour after the first lightning strike, the entire city plunged into darkness.

"The situation," an analyst concluded in Science, "could still have been saved by alert, well-trained operating personnel. They could, for example, have shed some load or increased generation to restore equilibrium. But Con Ed's control room succumbed to confusion and panic."[37] The room itself was antiquated. The modern control center for the New York power pool, in Albany, visualized in "a single dynamic display . . . the entire system: generation, transmission lines, transformers, relays, and breakers."[38] In contrast, Con Ed's operators could not easily grasp what was going on as they confronted a dizzying array of unintegrated meters, switches, diagrams, maps, and wall displays. Worse yet, the operator on duty did not want to take responsibility for cutting off customers ("shedding load"). He called his supervisor, who was at home, where the electricity was already off. The supervisor, with inadequate information, chose to reduce the voltage and to rely on a power line that neither he nor the operator knew was out of commission. Other managerial problems included inaccurate lists of the generators available on emergency standby and the failure in the previous months to repair a major grid connection to New Jersey, where additional power was available.

None of the many commissions looking into the blackout concluded that it met the legal definition of an "act of God." Nor did any of them find that it was caused by a lack of generating capacity or inadequate power in the grid. Rather, it resulted from numerous managerial mistakes and delays in maintenance. A federal report concluded that system collapse "resulted from a combination of natural events, equipment malfunctions, questionable system design features, and operating errors. Of paramount importance, however, was the lack of preparation for major emergencies such that operating personnel failed to use the facilities at hand to prevent a system-wide failure." A complete shutdown "could and should have been prevented by a timely increase in Con Edison's in-city generation or by manual load shedding."[39]

"To say that lightning caused the blackout," James Goodman concluded in his 2003 book *Blackout*, "would be like saying that the wind caused the capsizing of a poorly designed sailboat" with an incompetent crew.[40] The sense that Con Ed was incompetent led Ian Frazier to write a parody explanation of the causes of the blackout for *The New Yorker*: "8:37 PM. All is quiet at Indian Point No. 3 power station, when suddenly a huge dog jumps out of the bushes and eats several of the parts vital to operation of the plant's main generator." Further contributing to the failure, "every person in Queens between the ages of fourteen and thirty-six gets out of the shower and turns on a blow-dryer." In this whimsical "explanation," the system only buckled under when at "9:30 PM herds of buffalo, nobody has any idea where they came from, begin stampeding across New York State and start rubbing up against the exposed power lines where the dogs have eaten off the insulation."[41]

Consolidated Edison became the target of blame and parody because it was a "natural monopoly," with sole responsibility for the upkeep of the local electrical system. Because it was a vertically integrated enterprise, responsible for producing, generating, transmitting, and distributing power, Con Ed had no one else to blame for the failure except the Creator. But the blackout was not an "act of God." Consolidated Edison's cost cutting and technical compromises had weakened its system, and training and preparations were inadequate.[42] The reforms made after the 1965 blackout had been intended to make another major blackout impossible, but an improved grid and information sharing among utilities cannot make up for poor management.

Though 1977 is remembered as the year of blackout riots, most New Yorkers did not become instant criminals when the lights went out. Within three minutes, ordinary people, including many teenagers, had begun to direct traffic. Restaurants lighted candles and kept serving food as long as they could. Some actors attempted to go on with their Broadway shows, using flashlights. A man whistled at a woman and called out "Hey, beautiful"; she replied "How can you tell?"[43] For some, however, the blackout was immediately life threatening. On Roosevelt Island a special facility housed 44 quadriplegics who could live only with the aid of respirators, which had a battery reserve of four hours. After that, the patients might die. Well within that time, fire trucks arrived to generate the needed electricity. Neighbors assisted with flashlights and in other small ways. According to one resident, "the local community hung together beautifully."[44] At Columbia University, blind students served as guides in dark hallways and stairwells. Likewise, in some telephone booths the blind helped the sighted to place calls.[45]

When the power supply started to become irregular, alert dispatchers ordered subway trains to remain in stations. The blackout came shortly afterward, and most of the still-moving trains were able to coast into stations and discharge passengers. In the pitch darkness of the stations, it was difficult for people to find exits, but at least they were not trapped in the cars. In the entire system, only five trains were stranded underground. Two more stood on bridges. There was no panic among the passengers, who waited two hours or more in sweltering cars before being led to safety.[46]

As the blackout persisted, there were fewer automobiles on the city's streets—people able to leave the city had done so, and few ventured in. Instead, people who could not get home were on the streets and sidewalks, particularly in midtown Manhattan. Large crowds gathered in Times Square as theaters closed and subways emptied.

Because the power died after 9 p.m., many people were at home. Some looked out a window, stepped onto a balcony, or knocked on a neighbor's door to try to determine whether the blackout was widespread. Once people saw that power was out all over the city, many went back inside, lighted candles, and waited for developments. The managers of some apartment buildings knocked on doors to see if tenants were all right.

Everyone wanted to know why the blackout had happened, when it might end, and how large an area was affected. In the first minutes, however, not only did TV and radio have no answer to these questions; they were knocked off the air. The national TV networks went off for several minutes until their reserve power systems came on. When broadcasting resumed, the blackout could be watched elsewhere but not where it was taking place. The city's two all-news radio stations, WCBS and WINS, had backup generators and covered the story as it unfolded.[47] In the streets, people clustered around automobiles, listening to radios.

The tunnels under the Hudson River were closed, lacking ventilation, and commuter trains were stranded. The Port Authority Bus Terminal and Pennsylvania Station filled with people. "Large crowds of people wandered aimlessly around the area near Grand Central Station, wondering what to do next."[48] A bagpiper serenaded the captive audience of would-be passengers stranded in Grand Central Station's cavernous waiting room. At most hospitals, patients waited anxiously in the dark and were relieved when emergency power came on. But at Bellevue Hospital, the emergency power failed. Fortunately, no women were giving birth and no surgical operations were underway, but doctors and nurses had to hand squeeze air bags for the 15 patients on respirators.[49] At Brooklyn Jewish

Hospital, one woman gave birth by flashlight inside the building, small operations were conducted outside in the parking lot (with light and power drawn from generators on fire trucks), and at least 100 patients were treated, primarily for cuts from glass and knife wounds.[50]

Automobiles couldn't be retrieved from some garages and parking lots, because security systems were frozen and some gates or doors could not be opened. Guests of the Algonquin Hotel found that electronic locks had sealed their doors. The city's water pressure was not strong enough to reach above a building's fourth floor. Residents of high-rise apartments had to haul water up stairs in bottles and pails, use melting ice from their refrigerators, or, in a few apartments that had gas stoves, boil water from the toilet tank.[51] At the Windows on the World restaurant, 107 floors above the ground, diners were enjoying the famous view when the lights of the city went out. For three hours they continued to eat and drink amid gradually increasing distress. The toilets no longer worked. The kitchen's ventilation fans had stopped, and the restaurant gradually filled with smoke. But with no elevators working, the crowd of 600 kept the party going, got to know people at other tables, and kept ordering wine.[52]

In some parts of the city, however, conviviality, restraint, neighborly acts, and good Samaritans were scarce. A *New Yorker* writer who toured the city sensed a mixture of malaise, anger, and fear.[53] The Fire Department received thousands of alarm calls, the majority false. But there were more than 1,300 fires, 50 of them severe.[54] Looting was rampant. The police quickly arrested 3,750 people but soon ran out of cells to hold prisoners. Thousands more looters were never caught. In contrast with the racial riots of the 1960s, in which targets were selected, looting and burning were indiscriminate. The sense of citizenship and local community had broken down. A Ford Foundation study found that previous riots sparked by racial incidents did not provide a model for understanding this crowd's behavior. Race riots had begun with sheer anger and a desire to burn and destroy. Looting had not been the original impulse, but had emerged at a second stage. But during the blackout of 1977 there was no first stage; looting began immediately.[55]

At an electronics store at Broadway and Ninety-Ninth Street, looters began taking television sets, stereos, and appliances just after the power failed. (Clearly, their appetite for electrical devices was hardly dampened by the blackout.) When police arrived, the looters pelted them with bottles.[56] In Brooklyn's Bedford-Stuyvesant section and in parts of Queens, thousands of people surged into the streets almost as soon as the lights went out. They attacked neighborhood stores, starting with those selling sporting goods, jewelry, and electrical appliances and then moving on to clothing, furniture, and food stores. Because liquor stores were open and many of their owners were armed, "only" 51 were looted. Looters pried the steel gates off store windows by hand or yanked them off using trucks. The looters were black, white, and Hispanic, and they seldom distinguished between chain stores and independent businesses, nor did they spare shops that were locally owned, black-owned, or Hispanic-owned. But looters were discriminating about where they broke in, starting with shops selling luxury items that were easy to carry away. Police records show that more than 80 percent of those who were arrested first had criminal records. Most were men between 20 and 30 years of age, and they worked in teams. Often they prevented others from entering a store until they had taken what they wanted. Many teenagers and unemployed people under 35 joined in a second wave. An hour and a half after the blackout began, normally more law-abiding poor people joined in, including women and men who had jobs. Many people were "caught up by the near hysteria in the streets," as "social pressure to 'dig in' and take something was far greater than the normal pressure to abide by the law." These "stage-three looters" appeared just as the police were becoming more numerous on the streets, and many were arrested.[57] Nevertheless, whole neighborhoods were devastated. In one three-block stretch of Brooklyn's Utica Avenue (figure 4.2), nearly every store was looted.

New York was not dark during much of the 1977 "blackout," and the pillaging continued into the daylight hours. The word "blackout" had ceased to be closely related to light and darkness

Figure 4.2
An Associated Press photograph taken on Utica Avenue in Brooklyn after the 1977 looting.

and had come to signify the situation of lacking electricity. (Indeed, for years after 1977 "blackout" suggested a breakdown in the social order.[58]) The police usually did not know when a particular break-in began, and even when they knew they usually had difficulty getting to the site. In the early hours of the blackout, so many people crowded into the streets that they became impenetrable. Even at noon the next day, "roving bands of youth and adults were breaking into stores carrying off food, furniture, and television sets." At a Bronx automotive dealership, fifty new Pontiacs were hot-wired and driven away. Twenty-two were later found, stripped of their most valuable components. Others were found wrecked, probably by young thieves with little or no driving experience.

During the most intense looting, the police were vastly outnumbered, especially because officers were ordered to report not to their work precinct but to the precinct nearest to where they happened to be when the power failed. Areas experiencing little trouble (such as Staten Island) had extra police, while the places that most needed them (such as the Bronx) had too few. Nor was it any easier for police to move about during the blackout than it was for the rest of the population,[59] and it was impossible to follow the usual procedure of sending personnel from quiet areas to those in distress. Although the reduced ranks of the police could not restrain the crowds, fortunately they did not fire upon them. They were "expressly prohibited from firing to protect property."[60] Remarkably, there were few attacks on the police. Two officers who found themselves in a crowd of looters soon sensed that they were not in danger. "We were just pests," said one sergeant. "We were just something to get around so they could get at the goods."[61]

The newsweekly *Time* compared the riots to those that had occurred in 1968 after the assassination of Martin Luther King Jr. In an editorial, the *New York Times* agreed, declaring that the appropriate comparison was not to the 1965 blackout but to "the hot summers of the 1960s, which started in Harlem and Bedford-Stuyvesant in the summer of 1964, a year before Watts."[62] A Ford Foundation Study disputed these postmortems. It found that the looting was not a political protest but a "welfare disturbance."[63] Firsthand observers agreed. In their book *Blackout Looting*, Curvin and Porter assert that "the prevailing tone of the looters was non-political, and their purpose was to acquire items to fulfill personal needs or to sell to make money."[64] Of 1,328 stores attacked, the most frequent targets (48 percent) were stores that stocked clothing, food, furniture, and household appliances. This trend was confirmed by crowd behavior in drugstores, where people seemed less interested in drugs than in the clothing, food, small appliances, and other items on open shelves.[65] Sneakers were stolen more than anything else. One sociologist concluded that the "looting should not be thought of as a change in people but a change in opportunity."[66]

Twenty years later, the *New York Daily News* called the 1977 blackout "the worst night of the worst summer in the modern history of New York."[67] It called the looters "animals" and subscribed to the theory that a minority of low-life riff-raff had caused most of the trouble. If 1965 provided the backdrop for "a Jimmy Stewart moment," the *Daily News* went on, then the 1977 blackout was "the perfect soundtrack for David Berkowitz and his murderous siege." (Berkowitz, the serial killer better known as Son of Sam, was indeed killing people all that summer, but he remained in his apartment during the blackout.) For the *Daily*

News, which spoke for many ordinary people in the city, the problem was a breakdown in law, not hunger or poverty. In the Bronx, "looters struck an auto dealership and drove through the streets in a motorcade with horns blaring and pretty girls waving from the windows." These cars were stolen openly and displayed in an impromptu parade, not taken covertly and hidden away. The have-nots of society—"thousands upon thousands upon thousands"—"were out stealing," "many smiling."[68] In the *New York Times*, the labor historian Herbert Gutman commented on the general denunciation of the looters as "animals" and recalled the similar stigmatization of Jewish women who rioted over food prices in 1902.[69] This offended many readers, however, who in letters to the editor noted that the crowd in 1902 had not committed arson or indiscriminate looting, that they had been orderly, and that they had attacked only certain retailers who they believed charged excessive prices.[70] Few in the public accepted "the poverty explanation." A quarter-century later, a historian concluded that the "blackout of 1977 did not itself cause the economic trauma, crime hysteria, racial strife, and civic angst that suffused New York during this time, but it abetted and symbolized them."[71] The blackout both stimulated these forces and weakened the city's powers to thwart unrest.[72] The alienated and the disconnected saw in the blackout not only an economic opportunity but also an altered physical landscape that mirrored their everyday experience of being trapped. To them, the blackout emphasized the flaws of a poorly functioning metropolis, and unveiled the dystopian city for all to see.

David Mamet meditated on the 1977 blackout in *Power Outage*, a short play that was published in the *New York Times*.[73] A dialogue between two unnamed characters, it begins abruptly as one declares: "The thing which I'm telling you is no one enjoys being equal." This assertion contravenes the national ideology, announced in the Declaration of Independence, that all people are created equal. But illumination emphasizes social difference. The same character compares losing electricity to being in the locker room at the YMCA, "when you have taken off your clothes and they cannot see where you bought your watch." A brightly lighted society puts a premium on consumer goods and on achieving identity through attractive visibility. When the power fails, it provides ordinary people with the opportunity to seize the products they "need" in order to look good when the lights come back on. In Mamet's play, this is put in the form of the question "Why do we need these things?" The answer is "They keep us cool."

Mamet's play recalls Foucault's concern with the expression of power through mechanisms that, through the organization of space, regulate social behavior. By 1977 the United States had become a disciplinary society, as could be seen not only in formal institutions such as prisons, hospitals, asylums, schools, and military barracks, but also in the structures of everyday life. Thousands of electrical machines had created a dense overlay of controls on the movements of individuals, including traffic lights, subway turnstiles, automatic elevators, surveillance cameras, listening devices, motion sensors, burglar alarms, coded entry systems, and much more. All these systems of surveillance required electricity. Collectively, electrical devices established forms of control far more complex than mere visual observation, and created new power relations. But a blackout erased this disciplinary power, temporarily expanded the sense of freedom, and offered release into enhanced conviviality or riot.

When the power failed, New Yorkers did not necessarily embrace one another in Turner's shared *communitas*. The concept of liminality explained moments of ritual unity, but submergence in liminality was likely only if the citizens felt somewhat united before a blackout. They must have identified with their neighbors beforehand in order for the sense of *communitas* to emerge during a crisis. If cohesion is too weak, people will revert to Hobbesian competition. If the sense of community is strong, people will accept a temporary leveling and agree to share a common fate. When the sense of community has fragmented, some will likely reject a shared fate and seek to profit from the general disorder, to improve their position at least marginally before the lights come back on. During a blackout, people can choose either equality and solidarity or "devil-take-the-hindmost."

Fundamental to the shift in public behavior in 1977 was a change in how people perceived the city. It was not just more vulnerable; now it seemed potentially unsustainable. J. B. Jackson defines landscape as "a composition of man-made or man-modified spaces to serve as infrastructure or background for our collective existence."[74] Landscape is a shared creation. It is not something outside of human beings that they merely look at. Rather, landscape (including the city) is humanly modified space. Landscapes are neither natural nor the opposite of culture. They are the infrastructure of collective existence. To survive over long periods, people must find ways to live in and with their landscape. A landscape can suffer from intensive use, from the excessive application of chemicals and fertilizers, or from the failure to recycle. In the worst cases, landscapes may cease to support the people who live in them. When the power fails, the urban infrastructure becomes dysfunctional, and a long blackout may jeopardize the entire population. A city without electricity cannot sustain itself. Sewage backs up, food in freezers and refrigerators spoils, ovens and stoves remain cold, the water system fails, skyscrapers heat up to over 40°C, and the city becomes an unlivable heterotopia.

An anti-landscape is a man-modified space that once served as infrastructure for collective existence but that has ceased to do so, whether temporarily or long-term. Human beings can inhabit landscapes for generations, even millennia, but they cannot long inhabit anti-landscapes. This is not a new phenomenon. Archeologists have documented that some regions have been abandoned after being stripped of trees, overgrazed, or too intensively irrigated and farmed.[75] These were usually gradual processes. In contrast, highly technological societies can create anti-landscapes quickly, even suddenly. The nuclear contamination of Hanford, Washington and the chemical poisoning of Love Canal come to mind.[76] Most anti-landscapes are temporary, as when a fire involving toxic chemicals pollutes the air, forcing people to evacuate. A 24-hour electrical blackout is hardly as dire as what residents faced at Love Canal, where all residents were forcibly removed after it was officially declared toxic and uninhabitable. Yet a blackout temporarily transforms an urban environment into a space that cannot serve as the infrastructure of human existence. Instead, the infrastructure threatens to become a prison that people must escape to survive.

Each blackout is temporary, but the dysfunctionality of a society without electricity has become a latent condition. The networked city's life is inextricably linked to distant energy supplies and a vast infrastructure. Subtract electricity for more than a week from the networks sustaining American cities and suburbs, and they risk becoming uninhabitable. Without electricity, nearly all systems break down. Security alarms don't work, the water system fails, traffic lights cease to function, heating, ventilation, and air conditioning shut down, television and most radio go off the air, appliances become useless, computers crash, sewage piles up, gasoline pumps don't pump.

One way to mitigate these effects would be to redesign the electrical system. Rather than bind it more tightly together into a single integrated machine (the choice made after the 1965 blackout), decentralized electrical systems might be built that are not tightly coupled together. Generators separated from the larger grid could deliver power to essential services at hospitals, police, train stations, airports, tunnels, and highways. Such a plan was eventually implemented in California. But in the East, as Wade Roush has emphasized, "the major technological response to the blackout was a shoring-up of the existing power network rather than an attempt to supplement it with alternative sources."[77] Amory and Hunter Lovins, in their 1982 book *Brittle Power*, called for a more decentralized power system. They advocated power sources that would be diverse, local, and more autonomous than the existing grid, and they suggested that a dispersed system would be more resilient and easier to repair in a crisis. But the institutional response to both the 1977 blackout and the energy crises of the 1970s was to increase centralization.

The meaning of power failures had changed. In 1936, the electrical system had not yet been entirely assimilated into everyday life, and it was not taken to be "natural." By 1965, however, many New Yorkers regarded a blackout as a violation of the expected order of things. Yet it seemed an anomaly, without long-term implications, and the paralysis of that night became the occasion for a liminal moment. In later years, the perception of blackouts and the responses to them changed markedly. They became an annoying summer possibility, as the demands of air conditioning taxed the grid. They also became less potential carnivals than opportunities for criminality.

Historically speaking, this was not new. Urban darkness had a sinister air throughout the nineteenth century. According to Peter Baldwin, during the gaslight era, from 1819 until the 1880s, each time there was a blackout the newspaper stories focused on any disorder, immorality, or unrest that ensued. In particular, "warnings of crime and chaos grew strident in the early 1870s," when there were three gas blackouts in New York City. Repeatedly, "affluent observers feared that the order of the modern city would be snuffed out in an upheaval of hitherto suppressed social animosities and amoral impulses."[78] The actual experience of losing gas lighting seldom was as dire as these predictions. However, the perception remained that, although cities had been made safer and easier to traverse by the introduction of street lighting, they became vulnerable whenever the lights flickered out. Just as criminality later flourished in the shadow of the London Blitz, it could recur in an unintentional blackout. Darkening the city did not automatically unleash spontaneous harmony, as in 1965. It could also reveal a deeply fractured society that many New Yorkers, whether in 1977 or in 1872, feared could become a permanent condition. The blackout of 1965 had been an exceptional event that scarcely upset a confident and prosperous New York. But the blackout of 1977 appeared to signal civic breakdown and a grim future.

The looting left its mark on popular culture. An episode of *The Simpsons* that aired 25 years later depicts Homer and his family sweltering through a heat wave. The children rush to school because it is air conditioned, and an aerial view of the town shows its bulky air conditioning units sucking juice out of the grid. The power plant is already running at full capacity when Homer plugs in a dancing Santa Claus to evoke memories of winter. This small added demand pushes the system over the brink and plunges the city into darkness. Looting immediately breaks out, and Homer starts a security company that restores order so effectively that he angers the local mob.

Standing in contrast with the *Simpsons* episode are two films that were produced when the recollection of the 1977 looting was fresh. *Blackout* (1978) used the 1977 power failure as the background for a thriller in which four escaped prisoners terrorize a New York apartment building. Robert Carradine leads four escaped convicts in robbing one flat after another, opposed by only a single policeman, played by James Mitchum. The film depicts both the practical problems of living even briefly without electricity and the fears of rampant criminality. *Escape from New York* (1981) amplifies these fears. In it, the city has become a vast prison, inhabited only by criminals. When darkness falls, it is fatally dangerous to go into the unlighted streets.

The fear that only electrified security systems held civil society together reappeared in 1999, amid concern that computer systems would break down on December 31 at midnight. Would "Y2K" shut down electrical systems, close the stock market, imperil the water supplies, crash airline reservation systems, and lead to a long-term blackout? Again the un-electrified city was imagined as a terrifying, lawless place, a landscape of fear. A vision of New York in chaos and darkness also emerged in Billy Joel's song "Miami 2017 (Seen the Lights Go Out on Broadway)," which described a darkened Broadway in ruins. Nor were such disasters surprising, the lyrics explained; the apocalypse had been previewed in movies.[79] For a generation, Americans in general, and New Yorkers in particular, assumed that blackouts would lead to criminality and chaos.

To many journalists, the events of July 1977 seemed prophetic. The *Pittsburgh Press* editorialized that New York was "not the only place in America where riots and disasters bring out the ugly opportunists" and that "even in small towns, where people presumably are more considerate of each other, floods and tornadoes sometimes lead to larceny."[80] The *Idaho Statesman* argued that the New York riots were not atypical: "People are mugged, murdered, raped, and degraded daily. And it isn't just happening in New York. It's happening in cities across the country. Two-thirds of all serious crime is committed by children 10 to 17."[81] Many newspapers, including Vermont's *Burlington Free Press* and Florida's *St. Petersburg Times*, took up similar themes. The *Dayton Daily News* noted that the day the power failed the Census Bureau issued statistics documenting increased poverty. It condemned the looting, as did every other newspaper, including Harlem's *Amsterdam News*. But many editorialists argued that poverty and homelessness were contributing factors, and one wrote "What if there's ever a national blackout? 25.9 million of us are desperately poor. At last count."[82]

Few experts thought a national blackout imminent, however. After Consolidated Edison's many equipment failures and ineptitude became known, utilities elsewhere told their customers that a similar disaster was unlikely. British officials opined that their national grid, controlled from London, could meet any emergency, because it could generate 55,000 megawatts, one-third more than the highest demand ever recorded. Europeans felt equally certain that their multinational system, arranged more in concentric patterns than in lines, made it unlikely that a city could be as cut off from the grid as New York had been.[83]

A Los Angeles utility manager declared a comprehensive black-out a "very remote" possibility, and in San Francisco an official from Pacific Gas and Electric declared that a similar event was "highly unlikely because the systems were so different."[84] However, a California Utility Commissioner conceded that "brownouts on a rotating basis were possible this year if demand exceeded expected generation capacities."[85]

The 1977 blackout should have been a minor malfunction. It became a major outage because of faulty equipment and poor management. It was not due to fuel shortages or lack of generating capacity. Yet many people associated the power failure with the larger energy crisis. This misperception prepared them to confront rolling blackouts.[86]

During the late 1980s, the public discovered that blackouts, which once seemed merely temporary inconveniences that would be eliminated once local grids were fully integrated, were an inescapable feature of having an electrical system. Because the public resisted building new transmission lines and power plants, it became harder for utilities to meet demand. In the Northeast, generating capacity grew only one-third as quickly as consumption. In May 1989, the *New York Times* reported that "the East Coast from Maryland to Maine" faced "an unusually high chance of brownouts and blackouts on hot days . . . because peak electric demand is rising so fast that it threatens to outstrip supply."[1] In the same year, utilities resorted to rolling blackouts in Houston, Tampa, and Jacksonville, selectively cutting off groups of customers in rotation through their service areas. In 1988, Seattle suffered a three-day blackout during a heat wave that caused millions of dollars in losses. *Fortune* predicted widespread power brownouts in the 1990s and reminded readers that New England had experienced ten brownouts in 1988. The vice president of the US Council for Energy Awareness, a pro-nuclear-power organization, declared "Supply and demand curves in this business are approaching each other like a couple of express trains." *Fortune* further reported that "the people of Buenos Aires have been living with rolling blackouts lasting as long as six hours a day since December, the result of generator failures and a drought that has cut hydroelectric production."[2] The tempo of brownouts and rolling blackouts increased whenever temperatures approached 100°F.

A planned rolling blackout was quite unlike an accidental blackout. By definition, a rolling blackout lacked universality, affecting only a minority of customers at any one moment. It did not freeze time or paralyze society. It did not encourage the violation of social norms in a spontaneous eruption of either good feelings or criminality. A person caught in a rolling blackout knew that most of the system was still operating. The traffic lights still worked, the hospitals were open, and police and fire departments had power. Homeowners could minimize a rolling blackout's effects by not opening their freezers or refrigerators while it lasted, by securing water supplies in advance, and by purchasing batteries, candles, and other items. They could use a grill instead of an electric stove, light a few candles, and listen to a battery-operated radio rather than watch television. Since the rolling blackout usually lasted only an hour or two, it was an irritation to be worked around, not a confrontation with indeterminacy. In practical terms, the rolling blackout was preferable to letting the electrical system collapse, but it lacked the shared drama of unexpected darkness.

The "cure" for rolling blackouts seemed to be deregulation of the utilities, abolishing their "natural monopolies." From 1920 until 1975, Americans had believed that utilities that were regulated monopolies would be efficient, lowering rates for all. By the early 1970s, however, the possible economies of scale had been achieved, and growth in capacity no longer promised lower costs. After the energy crisis of those years, the old, inter-locking technical and economic justifications for utility monop-olies no longer seemed convincing, and consumers suspected electricity rates were unfair. Deregulation was widely expected to solve the demand problem, until in 2000 and 2001 Enron and other energy traders caused an epidemic of rolling blackouts as they swindled billions of dollars from California and forced several of its utilities into bankruptcy. This tale of illegal trading and unethical conduct suggested that the system would have been fine had energy traders been honest. It was not so simple. After both Enron and the utilities it had defrauded were bankrupt, rolling blackouts became less frequent but did not disappear.

During the energy crisis a generation earlier, President Carter, a trained engineer, knew better than to single out a scapegoat. Instead, he pushed Americans to take energy conservation seri-ously. Only a few parts of Carter's program became law, however, including a revision of the rules governing electrical utilities. The 1978 Public Utility Regulatory Policies Act (PURPA) under-mined the rate structures that Samuel Insull and his generation had created, which gave large firms lower rates. PURPA offered incentives to alternative energy production, and it sanctioned the rebirth of non-utility power generation, particularly as a by-product of manufacturing. Carter's legislation signaled that the political consensus that supported natural monopoly was crumbling. As Richard Hirsh notes, until PURPA the academic critics of the natural-monopoly doctrine could only present theories, but once it was in place "independent generators dem-onstrated in fact that non-utility players could produce electric-ity as cheaply as (or cheaper than) regulated power companies." They might achieve this by building cogeneration units that supplied both heat and light, or by installing improved gas turbines, adapted from aeronautical jet engines. When used in cogeneration facilities, they could achieve efficiencies well above 50 percent. By the mid 1990s, "they could beat utility costs by a margin of 5 to 15 percent."[3]

Before this challenge emerged, for generations local utilities had set rates low enough to lure and keep the large industrial customers that made possible economies of scale. A business producing its own power, such as a streetcar line or a shoe factory, tied up considerable capital in a powerhouse, and it needed generators with a capacity larger than peak demand. At night and on weekends and holidays, most of that generating capacity lay idle, and companies saw little point in owning expensive, underutilized equipment. They preferred to stay out of electricity production and focus on their core business. As long as a state utility board monitored prices and service, electrical monopolies seemed to make sense. But in the 1970s, the natural-monopoly consensus (which dated back to the presidency of Teddy Roosevelt) began to break up as electricity prices, after falling for 70 years, began to rise. Particularly after the emergence of new gas turbine technology, some large firms (notably food-processing companies) stopped being customers of utilities and became producers of electricity.[4]

Americans began to ask whether deregulation could hold down prices better than regulated monopoly could. Perhaps consumers could shop around for electricity, just as they did for other things. Most would be sensitive primarily to cost, but a consumer who wanted "green energy" from windmills or solar panels could buy that instead. Deregulation as a political philosophy also fit the mood of the country in the last quarter of the twentieth century. Even before Ronald Reagan pushed the idea during his presidency, Carter signed deregulation laws for airlines, trucking, and communications. Both Congress and the business community thought the sluggish economy would pick up if released from government controls.[5] The prices of airline tickets on heavily traveled routes did indeed go down, and in the 1980s telephone calls became cheaper after the AT&T monopoly was dismembered.

With these examples in mind, Americans began to implement the 1978 PURPA regulations in novel ways. While PURPA primarily dealt with electricity rate structures, the law somewhat unintentionally opened the door to more competition in the utility business. A decade after its passage, corporations that were not utilities were planning nearly one-fourth of new US power plant construction.[6] Presidents Ronald Reagan, George H. W. Bush, and Bill Clinton all treated energy as a problem that could be solved by stimulating supply through deregulation. They argued that competition could keep prices down more effectively than regulatory boards. Furthermore, technical aspects of the electrical system were changing as new communication and monitoring systems became available. The increasing sophistication of computers made it possible to shunt power from place to place more precisely and easily and to track transactions between many different producers and consumers. Control systems also improved, and utilities installed computers that simulated the grid and helped anticipate likely demand, relying on frequent sampling. The engineers working for a 1920s monopoly had lacked such capabilities.

Communications were also better. In 1920, long-distance telephone calls required operator assistance and did not always go through. The early telephone system was not fast enough for either utility emergencies or last-minute deals, much less monitoring. Within utilities, engineers might use telephones to discuss how to move electricity around, but there could be no regional and certainly no national market for the instantaneous sale of power. The first customer-dialed transcontinental call was not made until November 1951, when AT&T introduced area codes. Communication between utilities became more robust with the addition of e-mail and mobile phones. In the 1990s, deregulated inter-regional electrical markets made operational sense, in part because these high-speed communications could keep pace with the lightning movement of the electrical "goods" to be sold.

Advocates of deregulation argued that electrical lines were highways by which different suppliers could ship current. Instead of discrete islands of service, the entire country would be one energy marketplace. Electricity could move from buyer to seller at the speed of light. The assumptions of what Hirsh calls the "utility consensus" were discarded.[7] After the Energy Policy Act of 1992, many companies became exempt from utility regulation. Utilities were no longer permitted to be vertically integrated natural monopolies, and had to be split up. Some energy companies concentrated on wholesale production, others on the retail market. Lower prices, historically achieved through state regulation, would now be assured through competition. Efficiency, once achieved through the economies of scale, would emerge through competition and innovation. Market forces would set prices, assisted by traders, such as Enron. The public would treat electricity like any other commodity. The market would reward any scheme that could balance the load or generate power more cheaply. Electrification would be liberated from state control.

Even as the public became convinced that deregulation would solve the problem of supply through the magic of competition, the demand for electricity surged. In part, improvements in efficiency offset the gadget-hungry consumer's appetite. For example, refrigerators had become so much more efficient that the worst one on the market in 2004 used only one-third as much power as a standard appliance of 1970. During the same years, the country's energy intensity, or the amount of energy needed to produce a dollar of the gross domestic product, had declined by half.[8] Despite these gains, however, there was considerable truth in a cartoon showing a consumer, surrounded by an array of electrical appliances, making a phone call to ask what has caused the blackout.

By 2001, lighting accounted for only 8.8 percent of electricity use. More current was used on space heating (10.1 percent) or water heating (9.1 percent). The average household used as much power for its stove, dishwasher, and microwave oven (8.9 percent) as for lighting, and air conditioning (16 percent) or freezers and stoves (17.2 percent) used twice as much.[9] People bought more radios and stereos. They heated waterbeds and hot tubs. They acquired cable hookups, satellite dishes, video recorders, cordless phones, answering machines, humidifiers, dehumidifiers, power tools, and outdoor electric grills. Televisions grew larger and more numerous. Clothes dryers, rare before 1960, were found in 75 percent of American homes by 2001, and they used 5.6 percent of domestic power. Few Americans owned computers in 1975, but in 2007 they had 200 million, with accompanying printers and peripheral devices. Furthermore, the average size of a home kept increasing. There were more rooms to heat, cool, and light, and more spaces to fill with electric gadgets, such as exercise equipment and video games. Americans had forgotten there ever had been an energy crisis. The country's 107 million households averaged 10,650 kilowatt-hours a year, for a total of 1,140 billion kWh. Between 1940 and 2001 the average household had increased its electrical consumption 1,300 percent. A typical family used more energy every month than the grandparents had used in an entire year.

Yet few voices told consumers to cut back. When the anthropologist Laura Nadar studied energy experts in the 1970s,[10] she found, to her surprise, that engineers, scientists, and executives strongly believed that the best solution to the energy problem was to increase supply. They seldom discussed dampening demand. Virtually all the experts Nadar encountered equated technological progress with social progress. They assumed that certain trends—including increasing per-person energy consumption—were inexorable and irreversible. "The possibility that experts might be part of the problem was novel," and only toward the end of the decade did "the idea that the energy problem had human dimensions slowly begin to sink in."[11] Yet by the 1990s, for the first time, some utilities routinely could not meet summer peak demand and had to orchestrate brownouts and blackouts.[12]

When demand reaches capacity, an operator either "sheds load" or reduces voltage. Though electric lights can tolerate voltage reduction, some motors and most computers cannot. Therefore, utilities usually prefer not to play with the voltage level, and instead stop serving some customers. In sudden emergencies, the operators decide who will be cut off. But when the problem is foreseeable—excessive demand on a hot day, for example—the utility plans a "rolling blackout." New tools helped them cut off electricity selectively. Less than a year after the 1977 blackout, Lockheed Electronics received a patent for a "power monitoring and load shedding system" that could selectively connect and disconnect customers, using a digital computer with its own memory system.[13] During the following two decades other devices increased the ability of utilities to micromanage the load. A 1996 patent for an "energy management and building automation system" relied on microcomputers, installed both inside a customer's premises and on nearby utility polls, that were components of a comprehensive system for monitoring and controlling the flow of electricity.[14] If the rhetoric of deregulation promised endless supply, such technologies made it possible for utilities to micro-manage service interruption, as they divided consumers into groups to be blacked out in rotation. For example, Pacific Gas and Electric separated its customers into 14 service groups. During shortages, groups would be cut off in turn, each for an hour or two. During loss of service, customers were advised not to open freezers or refrigerators and to turn off appliances that might be harmed by a power surge when the blackout ended. Roughly 2 percent of all customers who provided essential public services were exempt from rolling blackouts, notably police and fire departments and hospitals.

Why couldn't utilities just build more capacity, as they had ever since 1880? In the early 1970s the means of generating power reached a technological plateau. Power stations had been getting larger and more efficient for nearly 100 years. Between 1950 and the early 1970s, the output of the largest generators had increased from 200 megawatts to as much as 1,300. But clever engineering could not increase efficiency further. Perhaps in theory steam heat and pressure could endlessly be increased, but metals reach limits when straining to contain superheated steam. The temperature limit for iron is 480°C, and that for austenitic steel is 560°C. Likewise, there are limits to how much energy can be extracted using techniques such as directing the same steam through a series of turbines operating at successively lower temperatures and pressures.[15] From 1880 until 1970, new power plants both increased capacity and lowered their production costs. After 1970, however, the cost of generating electricity increased inexorably for 15 years, until improved gas turbines became available.

Utilities also faced increasing public resistance to the construction of new power plants. Burning oil to make electricity had briefly boomed in the 1960s, but it became impractically expensive. By the 1970s, most of the promising sites for hydroelectric plants already had been developed. That meant new generation had to come from coal or nuclear plants. Because of smoke pollution and fears of radiation, however, few people would accept either a coal-powered or a nuclear plant "in their backyard." Attempts to build such plants were often thwarted by local residents in hearings before utility commissions and in the courts. In addition, millions of Americans were disturbed by the environmental effects of strip mining for coal. While France embraced atomic power, which eventually came to supply more than 80 percent of its electricity, grudging acceptance of nuclear plants in the United States gave way to fierce opposition after the Three Mile Island accident in 1979. Many planned nuclear facilities were never completed, and as a result atomic reactors provide only about 20 percent of US power.

Another promising alternative seemed to be burning natural gas, but during the 1970s its price increased as it became more popular. In short, the technical and economic situation for producing electricity changed fundamentally. Sixty years of declining electricity prices were over. Deregulation could neither increase the efficiency of generators, nor create new dam sites, nor raise the melting point of steel. Nor could deregulation convince the public that nuclear plants were desirable or that strip mining Appalachia was a good idea. Nevertheless, in the mid 1990s deregulation of electricity was ballyhooed as a sure way to create more power at lower prices. Leading the chorus of lobbyists for this change was Enron, which began to buy and sell electricity.

It did not take long to test deregulation in practice. California's deregulation law was adopted in 1996, when demand for electricity in that state was flat and generating capacity exceeded demand. It went into effect in March 1998. Most experts agreed that costly investments in nuclear plants and alternative energy had made Californian electricity expensive, but even without deregulation this situation was expected to improve because of (expected) lower gas prices and improved depreciation write-offs for nuclear plants.[16] The utilities themselves supported a deregulated market, and quickly divested themselves of most of their generating capacity. Pacific Gas and Electric had spent a century building a diverse system that supplied most of its customers' needs, but after selling off its fossil fuel plants and retaining its hydro-electric dams it could supply only about half as much electricity as its customers demanded.[17]

California utilities had installed 760 megawatts of alternative energy production before deregulation, but afterwards they built no additional generating plants of any kind, even though both the state's population and the demand for electricity increased. Yet during the first year of deregulation California's wholesale electricity prices fell, although the market was more volatile than in the past. Enron and the other energy traders needed and fostered volatility in order to thrive.[18] Yet overall, it briefly seemed that deregulation was working. In contrast, during the summer of 1998, rolling blackouts were necessary in Colorado and Chicago, strengthening the case there for an end to monopoly.[19] The following summer 100,000 Chicagoans suffered a blackout on July 30—the hottest day of the year, at 104°F. Two weeks later a substation malfunction shut down the entire central business district.[20] Briefly, Californians and champions of deregulation felt justified. But then the situation deteriorated on the Pacific coast, which had less rain than usual. This reduced the hydropower available in California and Oregon, where dams on the Columbia River were important suppliers. By 2000 dwindling water supplies were inadequate to deal with a May heat wave. During California's summer and early fall, the *average* price of a megawatt of electricity was quadruple the 1998 price. By Christmas, the *average* wholesale price had soared to ten times that of two years before. In early 2001, the average was over $300 per megawatt-hour. The major utilities could not afford to buy at these inflated wholesale prices, nor could they prevent traders from creating artificial shortages, and rolling blackouts became frequent. In January, California's state government intervened in the market, buying power and allowing retail prices to rise. Nevertheless, in April 2001 Pacific, Gas and Electric—one of the largest utilities in the United States—filed for bankruptcy protection.

What had gone wrong? From Enron records, transcripts of trader conversations, and interviews with regulators, the *Wall Street Journal* concluded that the "energy companies seized on loopholes [in the new deregulation laws] and local shortages to charge prices hundreds of times higher than normal."[21] Or, as a prominent lawyer dealing with the electrical industry put it, "merchant generators, using the utilities' divested assets, could largely control the price of power in the day-ahead market outside the purview of state regulation."[22] At times, Enron pushed up prices by shutting down plants or prolonging repairs and maintenance to create artificial shortages. When prices had been regulated, a megawatt-hour often cost less than $10 on the wholesale market, though at times the price was above $30, and at peak it might rise to over $100. After deregulation, a megawatt-hour of stand-by power might sell for more than $700, and at least on one occasion the price was $9,999. Eventually, the Independent System Operator (ISO) exchange imposed a price cap of $250 per MWh, which still was grossly higher than the cost of generation. During the decades of regulation a utility had to prove that prices bore a clear relation to production costs, but deregulators had embraced a naive belief in the benign, invisible hand of the market.

But malign (and eventually visible) hands were soon in control. Deregulation made it more profitable to sell small amounts of power at high prices than to provide a full supply. The new generator owners discovered that it could be more profitable to close some plants for maintenance than to keep them all operating.[23] At times during 2000, as much as one-fourth of California's power generation was shut down, five times more than was normal before deregulation. In 2001 shutdowns occasionally included almost one-third of California's total capacity.[24] Another egregious practice was Enron's overbooking of transmission lines, in order to receive a fee of up to $600 per MWh for giving up transmission rights.[25] As they relieved the fictive congestion they had caused, traders chatting on the telephone called these scams "Fat Boy" and "Get Shorty."

In another technique, nicknamed "ricochet," traders contracted to sell Californian power to out-of-state middlemen, creating a shortage that drove prices up, then re-sold the same power back to California at a much-inflated price.[26] Suppliers, particularly gas companies, also profiteered by creating shortages. By the end of 2000, natural gas prices in California were 600 percent higher than in the rest of the United States.[27] There was a widespread pattern of collusion among energy suppliers. As a lawyer familiar with the case concluded, "where a single company [such as Enron] has significant concurrent interests in both gas and electricity, the opportunity for market manipulation expands exponentially."[28]

State investigators concluded that the energy traders overcharged Californians by $9 billion, and that not Enron but other corporations had the largest profit increases: West Coast Power (745 percent), Dynegy (230 percent), AES (181 percent), and Duke (110 percent).[29] The utilities that these traders cheated on the wholesale market could seldom recoup their expenses by charging retail consumers more. In fact, California law mandated a 10 percent cut in retail rates so that customers would be certain to share in the "savings" expected from deregulation.[30] The only way a utility could avoid this mandate was if it had eliminated all of its "stranded costs." Stranded costs result when a utility suffers losses as a result of legislation. For example, a utility might plan to pay for a power plant on the basis of a rate structure that disappears with deregulation, or it might have to close a nuclear plant to comply with new environmental laws. These stranded costs became a tax write-off. Those Californian utilities that had such stranded costs (and the accompanying tax break) had to reduce retail prices 10 percent. They were forced to sell power for as little as one-fifth of what they paid for it, thereby losing millions of dollars every day.[31] Previously profitable companies went into bankruptcy. San Diego's utility was one of the few exceptions. It escaped low retail rates by paying off all of its stranded costs, and therefore could double residential electricity bills between May and August. It remained solvent, but its consumers were enraged.

As Enron and other traders created artificial shortages, rolling blackouts darkened Californian communities and periodically shut down Silicon Valley. In 1998 and 1999 California had not experienced even one "stage three power emergency" (the bureaucratic term for a rolling blackout). These began if power reserves fell below 1.5 percent. But during the first 150 days of 2001 Californians experienced 38 rolling blackouts.[32] On January 17, for example, blackouts disrupted 1.8 million customers from Bakersfield all the way to the Oregon border. On March 19, the California system as a whole was 500 MW short and rolling blackouts affected half a million customers. The public was furious, and many felt the state's years of rapid growth might be coming to an end. A *New Yorker* cartoon (reproduced here as figure 5.1) suggested that the grandchildren of the "Okies" who had left Oklahoma in the 1930s might return to that state, where electricity was plentiful.

Public institutions were supposed to be exempt from rolling blackouts, but during these 38 days of blackouts several schools and two hospitals lost power, and some traffic lights stopped working.[33] Universities were not spared. In 1998, the University of California and California State systems had negotiated a statewide four-year contract with Enron. California State Polytechnic University at Pomona had an "interruptible" contract that required it to curb demand drastically during shortages or face penalty charges that could reach $50,000 an hour, and

"Get in, Tom—we're goin' to Oklahoma. Hear tell they got 'lectricity."

Figure 5.1

A cartoon published in the February 19, 2001 issue of *The New Yorker* suggests reverse migration out of California, back to the former Dust Bowl.

many other universities had similar contracts. On blackout days, most campuses shut down. The Claremont Colleges had ten blackouts in a single week, totaling 41 hours. In the last week of one term, because of examinations, those colleges decided not to shut down but to reduce power use. They were charged a penalty of $257,000.[34]

The costs were even greater in the dairy industry. Cows must be milked twice a day or they become ill. Farmers use electricity to power milking machines, to cool the milk, and to heat water for sterilizing equipment. And without electricity the largest milk plant in the United States, run by Land o' Lakes, could not process the 12 million pounds of milk that were trucked in every 24 hours. During the 1990s the power was never interrupted, but from mid December until mid January of 2001 it was off 17 times, and tons of milk were dumped rather than processed.[35]

When a computerized company loses power, unsaved data is lost, and some of the more delicate equipment may be damaged by power fluctuations or surges. "Every time our system crashes," a California oil pipeline manufacturer complained, "it corrupts a lot of data files, and we have to go to tapes of the day prior, and everything done that day could be lost."[36] In response, the company spent $40,000 on a backup battery system that could sustain operations for one hour, long enough to save data and organize a controlled shut down. Disruption and loss were much greater in software companies. Indeed, nearly all of California's businesses lost money as a result of the rolling blackouts. Many businesses began to install backup power systems.

After witnessing the California debacle, many states put deregulation plans on hold, and utilities faced an uncertain future.[37] Many utilities operated in several states, some deregulated and others not. Left in a muddle, they were cautious about investing in power plants or distribution lines before the legal situation clarified, and thus deregulation led to stagnation in infrastructure construction. Deregulation had worked for the telephone industry and the airline industry, but it had failed to provide cheaper power. At times, it failed to provide power at all. The US Department of Energy found that in the period 1999–2004 there were more than 125 "significant" blackouts—roughly one every two weeks. How would Americans have felt about airline deregulation if two planes had crashed per month? Could this debacle be blamed on a poorly organized electricity market, or was it due to the greed of corporate traders, particularly Reliant, Duke, Williams, and Enron, who "gamed the system"? Were electricity markets, which sell something that cannot easily be stored and must be consumed as it is produced, fundamentally unlike the market for pork bellies or that for gold?

Comparisons with air traffic control and the telephone system help explain why it was difficult to deregulate electrical utilities. The federal government operates a radar network that blankets the United States, tracking all planes in the air or preparing to take off. In contrast, the electric power industry of the 1990s could not track in real time the movement of current over a national system of high-voltage lines. The sheer speed of electrical transmission made real-time monitoring difficult. Imagine how hard air traffic control would be if planes moved faster than radar signals. Utilities often relied on computers to extrapolate from sample data to model what was likely to be happening on their portion of the grid. This worked most of the time, just as relying on airplanes to follow their flight plans worked most of the time. But not all planes stay precisely on course, and sometimes one crashes as a result. That is why the federal government created a comprehensive radar system in the 1960s and made all airlines subject to the same air traffic control regulations. The electrical grid was not monitored by a comparable federal agency, even though real-time monitors are available that could make the system more like air traffic control. "If the existing 157,000 miles of transmission lines in the United States were fitted with $25,000 sensors every 10 miles, and each sensor were replaced every five years, the annual cost would be $100 million."[38] This might seem a large sum, but it is less than 3 percent of the cost of the 2003 Northeast blackout, and it would only add 0.004 cent per kilowatt-hour to the average consumer's electric bill.

Airlines and utilities have other problems besides knowing where the planes are or where the electricity is. Both can experience excessive demand. Instead of having extra planes or additional power plants, they can send passengers on alternative flights or buy power from other suppliers. This is where deregulation might have a beneficial effect. Instead of every utility owning expensive excess capacity, a few strategically placed plants could provide it. Here again, transmission-line sensors could give precise information that could be used to optimize performance.

The need for national systems of coordination arose at roughly the same time for the airline industry and the electrical industry. Initially, both relied on corporations to set standards. But after some spectacular plane crashes in the 1960s, voluntary industry-wide organizations gave way to stringent federal standards, mandatory training, and inspection. The airline industry is deregulated, but its firms must comply with the federal regulation of air traffic control. After uniform standards were imposed and a national radar system was established, airline accidents plummeted. Yet it remained acceptable for utilities to fail every customer for a total of 24 hours over a ten-year period. It would not be acceptable for each airline passenger to experience one crash per decade. After deregulation, electrical utilities still relied on voluntary compliance with standards suggested by the industry. It would have been wiser to mandate compliance and real-time monitoring as preconditions to deregulation.

A second comparison further explains why there were difficulties in deregulating US electrical utilities. The telephone system also was long organized as a natural monopoly, and the decision in the 1980s to break up AT&T and to allow competing companies to emerge was an important precursor to breaking up the power companies. But there were crucial differences between the telephone system and the electrical system. The most important difference was that competing telephone companies could send messages down many paths. The traditional telephone lines that hung along roadsides had been supplemented by satellite relays and fiber-optic cables. The telephone system was robust, and it had excess capacity, even before the Internet made further options available. In contrast, there were fewer high-transmission electrical lines. Sometimes there was only one practical way to connect a particular supplier to a particular buyer. This obviously inhibited full competition. Furthermore, when electrical systems were deregulated, this weakness was not immediately addressed. Instead, as the official report on the 2003 blackout concluded, "two decades of electricity policy reform measures have focused on inducing investment in generation to reduce or avoid the need for grid expansion."[39] One utility executive told a conference at Harvard University that insufficient transmission had protected local power markets, inhibited competition, restricted customer choice, and threatened the reliability of the system.[40] The result was a power glut in some locations, a power famine in other areas, and inadequate lines to move the energy to where it was needed.

The electrical supply system also differed from telephone networks in another important way. Telephone service had been intentionally built into a national system, but electricity distribution had been organized into largely self-contained regional systems (the East, the West, and Texas), with three main inter-

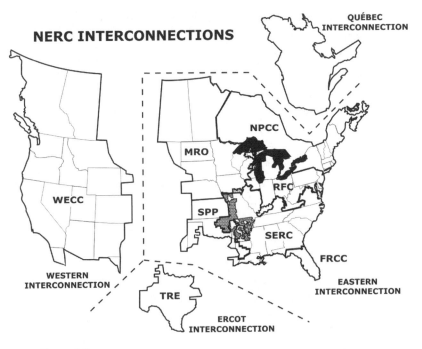

Figure 5.2
Divisions of the power grid in the United States. Note that the Western Interconnection is not subdivided and includes one-third of the United States, and that Texas is a separate unit. Courtesy of North American Electric Reliability Council.

connections. (See figure 5.2.) Dialing and switching technologies made it easy for millions of telephone customers to tap directly into the national system. Electricity consumers were in a far different situation. They had no direct access to different utilities, and there was no comparable national grid. Anyone, using any phone, can dial a call to Kennebunkport, Houston, or Sacramento. But no local utility, much less any individual homeowner, has direct access to power from these three widely separated places. Each is in a different regional system, with only limited lines of interconnection.

When power monopolies disappeared, operating electrical grids became more complex. Before, the goal had been to produce only what was needed nearby, with a little excess in spinning reserves. But a deregulated market encouraged construction of power plants that could produce much more than was needed locally. A deregulated market also included trading companies, such as Enron, which generated little or no power and owned no transmission lines. When trucking was deregulated, the national government had already constructed a vast interstate highway system, supervised by state police. But a program of erecting transmission lines did not precede or accompany utility deregulation. Rather, between 1975 and 2005 utilities reduced investment in those lines by 25 percent.[41] After the mid 1990s, utilities had little incentive to construct high-voltage lines, because the transmission business was being detached from generation. The economist Paul Joskow warned in 1999 that "proceeding on the assumption that . . . the market will provide needed network transmission enhancements is the road to ruin."[42] In practice, utilities often delayed replacement of transmission equipment. These "economies" hindered power transshipment. "The number of times that the transmission grid was unable to transmit power for which a transaction had been contracted jumped from 50 in 1997 to 1,494 in 2002."[43] With nearly five delivery failures a day, was there really a free market? In 2001, failed transmissions and congestion cost more than $800 million, according to the Federal Energy Regulatory Commission (FERC). The following year, the PJM system (serving much of Pennsylvania, New Jersey, Maryland, and Virginia) had congestion costs of $430 million, and New York's Independent System Operator had congestion costs of $525 million.[44]

One might expect an industry with such problems to spend more on research and development. But another effect of deregulation was to reduce utility companies' contributions to the Electric Power Research Institute, their chief R&D organization.[45] Indeed, both public and private investment in research and development declined from its peak in the 1980s to 2005.[46] Utilities were even reluctant to spend money on proven technologies that could maximize use of existing equipment. In 2007, fewer than 100 transmission lines in the national network used "dynamic thermal circuit rating" (which helps transmission operators to see "the actual rating of a line," an indication of how much additional power it can handle).[47] Instead, the operators had to estimate.

Not only did deregulation discourage utilities from enlarging transmission capacity; consumers rejected high-tension lines in their neighborhoods. In 1979 a controversy arose about the safety of such lines. Studies made in Denver had found a higher incidence of cancer among children who lived close to high-tension lines. Though subsequent research had difficulty confirming the correlation or tracing causation, the public did not want to take chances.[48] As scientific debate continued, homes near transmission lines decreased in value, lawsuits multiplied, and public resistance to new lines intensified. Even without such opposition, and even if new transmission lines were not delayed by cost cutting, utilities often did not want to build them. As one expert put it, why should a local utility "endure the political, financial, and land-use costs of siting local grid upgrades"[49] for the dubious reward of increasing a competitor's access to its market? For all these reasons, investment in new lines stagnated.

By 2000, the combination of increasing demand for power, lack of new power plant construction, decades of underinvestment in transmission lines, and inadequate monitoring systems pressed the system to its limits. By 2006, "expensive plants built just to meet peak demand, as on hot summer days" were being "run 40 percent of the year."[50] Deregulation could not instantly fix problems that had been built into the system over decades. The grid was ripe for malfunction, even without Enron's rapacity. In 2003, former Secretary of Energy Bill Richardson declared: "We are a major superpower with a Third World electric grid. Our grid is antiquated. It needs serious renovation."[51]

Richardson was referring to a blackout that occurred on August 14, 2003, affecting the area from Michigan and Ohio through Toronto to New York City. It cost consumers an estimated $7 billion. The official joint report, issued by a Canadian-American task force, concluded that the First-Energy Corporation, headquartered in Akron, was primarily responsible. First-Energy had failed to fix inadequacies in its system. It had neglected tree trimming in its transmission corridors, in some areas for as long as 14 years. It had not made effective accident contingency plans. The report concluded that the blackout would have been entirely preventable had First-Energy and other utilities adhered to the non-binding standards laid down by the North American Electric Reliability Council (NERC).[52] But without mandatory inspections, and with no fines for failure to comply, the grid as a whole was only as strong as its weakest, non-complying link. The report pointed to human error as the blackout's cause, but gross negligence would have been more accurate.[53] NERC concluded in 2004 that "market-related needs are causing the grid to be operated closer to its reliability limits more of the time."[54] Monitoring also remained difficult. There was inadequate inter-regional visibility over the power system, and on that particular afternoon in 2003 utilities in the Northeast did not know what was happening at First-Energy because its computer system was off line for repairs. The blackout was what one could expect from an under-regulated utility system relying on outmoded monitoring equipment and inadequate transmission lines to meet rising demand.

NERC was created in 1968, when each utility had a monopoly over generation, transmission, and customer service in its geographical area. Each was responsible for a largely self-contained local system. Each had no one else to sue or to blame for a blackout. Before deregulation each incurred the losses if its part of the grid failed. Once utilities ceased to be monopolies, however, they were not responsible for the system as a whole, and they had weaker incentives to follow NERC's voluntary guidelines. Moreover, creating contingency plans became problematic because it required collaboration with competitors. In the new business environment, NERC watched helplessly as companies flouted guidelines. "If we had had a solid compliance program in place to see to it that rules were being followed," its president declared, "we would not have had this blackout."[55] The organization finally received more authority, and with it a slight name change that did not affect the acronym NERC. After June 4, 2007, it had authority to enforce 83 detailed standards, and its own operations were monitored. These standards required that utilities have detailed and complete real-time data on load balancing and the interchange of power, that they continuously staffed voice and data communications (to be logged or recorded for later study if required), and that they report on all plans to shed load.[56] In addition, any operator responsible for part of the interconnected transmission system had to take part in annual simulations of emergencies and extreme conditions.

Yet if standards and enforcement were in place in 2007, transmission-line budgets were insufficient to build the energy interconnections required to create a deregulated marketplace. In many areas, notably California, the grid was stretched to the limit. It was also aging. The US Department of Energy noted in 2003 that "most existing infrastructure—wires, transformers, substations, and switchyards—has been in use for 20 years, or more."[57] The volume of transmission transactions increased 400 percent between 1990 and 2004, but high-voltage lines expanded by only 0.2 percent a year.[58] An expert warned: "Rising demands and continuing low grid investment remain on a collision course."[59] After the 2003 blackout, NERC estimated that the cost of upgrading the country's transmission lines would be $56 billion, but by 2008 the DOE's estimate was $100 billion.[60] Far less was being spent. As *Business Week*'s headline summarized, it was time to "stop bickering and fix the grid."[61]

Even if energy traders had not gamed the system in 2000–01, there were many potential causes for blackouts. The public resisted building new generating capacity, yet continually demanded more electricity. Private utilities avoided building transmission lines, skimped on maintenance, and resisted mandatory federal standards for their operations, preferring voluntary adherence. At the same time, by championing deregulation the federal government undercut state regulation agencies, without first ensuring construction of an adequate grid or comprehensive monitoring. This left computers to simulate—and utility operators to estimate—the movements of electricity. This task became even harder as the private energy traders created artificial shortages by shutting down generating plants and as they booked movement of electricity over the lines in excess of what was possible in order to profit from the resulting congestion. Rolling blackouts were often the fault of energy traders, but they exploited regulatory loopholes and pre-existing technical weaknesses.

An unwieldy amalgam of public and private actors had built inconsistencies into the electrical system. The public resisted building new generation and transmission facilities, yet supported deregulation that would work only if there were a surplus of power. Deregulation discouraged investment in the grid and encouraged cost cutting on maintenance. Utilities reduced investment in transmission lines and resisted comprehensive monitoring. Rolling blackouts were hard to avoid when deregulation took place under such conditions. The pretense that the largest machine in North America was a marketplace prevented the government, private enterprise, or the general public from understanding that electricity is not simply a commodity that moves to a market. Rather, electricity is the instantaneous product of a vast interlocking machine whose parts must be synchronized at all times to produce a uniform alternating current. A sudden change in either supply or demand in New York can disturb the machine all the way to Detroit. A surge in Fresno can cause a blackout in Montana.[62] The whole system remains inextricably interlinked, from the steam turbines and hydroelectric dams at one end through high-tension lines, substations, distribution lines, and meters to the consumer's wall plug.

Because electricity is produced a fraction of a second before being consumed, it is difficult to divide wholesale from retail, or production from distribution. Unlike almost any other commodity, electricity cannot be extracted from the vast mechanism of its production and stored for later use. It is always flowing through the system, and not necessarily following the shortest route between two points. When a system of high-tension lines first encircled Lake Erie, system operators soon observed "very large circulating power flows around the lake [that] occasionally reached 1,000 MW."[63] Sometimes they moved clockwise, sometimes counterclockwise. Similar flows circle the Rocky Mountains. Carving up a mechanism that can behave in such unpredictable ways divides the responsibilities and the economic interests of generators, wholesalers, traders, transmitters, and retailers. Increasing the number of operators complicates coordination, and it heightens the chance of malfunction or blackout even if no one is "gaming the system." In a Senate hearing after the 2003 blackout, the CEO of the PJM Interconnection said: "We are in the middle of a long and difficult transition. We are dealing with a speed-of-light product that does not respect state or even international borders. Yet, this industry was built, financed, and operated for over 80 years as a gaggle of over 4,000 different entities. . . ."[64] Coordination could not be left to "the invisible hand" of the market because electricity ultimately is not a product like bananas sitting on a counter waiting for a buyer, but the fluid and evanescent expression of a vast mechanism.

Congress finally acted to improve electrical grids with the Energy Transmission Act of 2005. Among that act's many provisions was the requirement for regular analysis of transmission congestion, mapping energy corridors (for oil, gas, and electricity) and designating "national interest electric transmission corridors." The latter can exercise the power of eminent domain if that seems necessary. Two "national interest electric transmission corridors" were soon identified. One runs from Northern Virginia to Albany, New York; the other encompasses greater Los Angeles and stretches over the mountains past the dams on the Colorado River into western Arizona. The word "corridor" misleadingly suggests a tightly focused strip of land. In fact, the eastern "corridor" includes all of New Jersey, Delaware, and Maryland and more than half of Pennsylvania, West Virginia, and New York. In these areas, the federal government can now push energy transmission in order to modernize existing equipment and expand capacity. However, it will take years to overhaul and expand the electrical grids, because the Department of Energy must identify problems, FERC must approve technical solutions, and all projects must pass environmental impact assessments, which usually are stringent when under federal supervision.[65]

Meanwhile, rolling blackouts never went away. Nevada had its first in 2001,[66] and there were others during the following decade. For example, in 2006 rolling blackouts were necessary in Colorado (March), in Texas (April), and in California (July), and they were narrowly averted in Massachusetts (August).[67] After no new plants came on line for two years, the *New York Times* concluded that "parts of New England could face Third-World-style blackouts in coming summers if demand for electricity, particularly for air conditioning and refrigeration, exceeds available demand."[68] In 2008, warnings of rolling blackouts were issued for Texas, Ohio, Massachusetts, and New Mexico. Unplanned blackouts also remained common. As Joskow noted, many smaller blackouts in these years, notably in New York, were not caused by generation shortages but by equipment failures, often because local transmission systems were aging.[69]

Blackouts also occurred in other countries, notably Canada (July 2006), Indonesia (July 2008), and South Africa (November 2007 through March 2008). Yet the difficulties of the American electrical industry are not identical to those elsewhere. The United States experiences a blackout somewhere every 13 days, and the frequency has not changed greatly in the last 30 years.[70] Most European countries experience fewer power failures. Perhaps the most striking contrast is with Russia, which also has an enormous grid spanning a continent. After the collapse of the USSR and the economic difficulties of the 1990s, one might expect to find a corresponding disintegration of the Russian electrical system. In fact, however, between 1975 and 2005 the Russian system did not suffer many significant blackouts.[71] Its hierarchical organization permitted administrators to insist on a centralized power dispatch system, automatic emergency control systems, coordinated maintenance scheduling, regular inspections, multi-level backup capabilities, and transparency in operations. The Russian grid as a whole is monitored constantly, using computer modeling, load forecasting algorithms, and other diagnostic tools. The information is displayed on "visual walls" that provide "a comprehensive, easy-to-understand, changing image that helps a real-time dispatcher quickly apprehend and understand the situation."[72] In such a control system, cascades "are normally arrested" in the "very early stages by the energy control systems."[73] Likewise, the more centralized parts of the US electrical grid—notably PJM in Pennsylvania, Virginia, New Jersey, Maryland, and other contiguous areas—successfully weathered the 2003 blackout. In contrast, there were two major blackouts in 1996 inside the more loosely organized western system, where some companies resisted NERC's voluntary controls.[74] Much of the US electrical grid is not sufficiently transparent to those who operate it, because it is splintered into many units and subdivided into electricity wholesalers, distributors, and retailers.

The grids in most European countries do not produce as many blackouts. Nevertheless, large-scale power failures have become a notable feature of all electrical grids. Just two weeks after the US-Canada blackout in 2003, London suffered a power failure. Four weeks later, another struck Sweden and Denmark. One week after that, all of Italy lost power. In May 2005, Moscow suffered one. In each of these cases, the systems seemed to be operating normally, and there was little to suggest that a major blackout was about to occur. But in each case unscheduled line losses and other equipment failures stressed the electrical systems and made them vulnerable. Next, a "triggering event" occurred. Typically one transmission line tripped out, causing surges in power to other lines, creating overloading and irregular voltages. These problems then prompted some equipment to shut down automatically or to cut off from the rest of the grid. The surges and overloading increased, leading to a cascade that shut down the entire system.

Some scientists have analyzed these events. Early studies tended to be retrospective, focusing on technical difficulties or human errors that led to a particular blackout. In recent years, a more theoretical approach has become common. Blackouts, whether caused by poor system design, excessive demand, human error, bad weather, or technical malfunctions, can be modeled, simulated, and statistically analyzed. Most researchers seem to agree that large-scale blackouts are unavoidable. John Doyle, a specialist in control theory, developed a mathematical model that matched the historical statistics for power-system failures. In North America, between 1984 and 2000, eleven blackouts occurred that involved the loss of 4,000 MW or more. That was one every 18 months, far more often than would be predicted by extrapolating from the declining frequency of small blackouts. Why should a large, complex system that is vigilantly watched day and night be robust most of the time, yet occasionally prove fragile? Doyle explained the frequency of power outages in an optimization model. It posited that utility engineers make rational choices, focusing on improvements and maintenance in their local system, and that they anticipate problems and generally prevent cascading power failures. But in making their technologies more robust and interconnected, they also increase the grid's complexity, making it more vulnerable. The largest blackouts are difficult to anticipate in good part because of this sheer complexity, which makes the grid curiously fragile in certain circumstances. The failure of one circuit breaker, substation, or transmission line can set in motion an unforeseeable cascade of effects, as electricity oscillates through the system in unpredictable patterns moving at the speed of light.

Researchers at Oak Ridge National Laboratory have developed a second explanation for why small outages have been reduced more effectively than large ones. They regard blackouts as feedback loops that, over the long term, spur improvements in the system. They assume that utilities try to deliver power as cheaply as possible, maximizing profits, until they push the system too far, causing a major failure. Each outage forces them to improve transmission lines, relays, and computerized controls. Thus, a blackout is "feedback" that counterbalances the managerial drive for profits. Far from being a purely negative event, it has the positive function of forcing upgrades. These researchers have also simulated the behavior of power networks and found the results compatible with NERC statistics for 15 years of actual blackouts.[75]

Whatever their explanations, all researchers recognize that blackouts do not simply become more frequent as a function of the intensity of demand, nor do blackouts necessarily become larger because there are more interconnections. Nor does a cascading electrical failure gradually become more probable as a result of a straightforward cause-and-effect relationship. Rather, "there is a transition point at which its likelihood sharply increases."[76] Just as water's likelihood of turning into steam does not gradually increase with temperature, but changes radically at 100°C, the chance that a blackout will occur passes through a "point of criticality."[77] To determine what factors shape criticality, one group constructed a simulation model consisting of 150 generating units and 1,800 transmission lines and transformers. The simulations discovered a critical loading point, beyond which the system had far less resilience should any component fail. Once a system passed that point, there was "a sharp rise in mean blackout size and a power law in the probability distribution of blackout size."[78] This might suggest that an electrical system should have more technical redundancy and more excess capacity, but when making risk assessment "there are substantial economic benefits in maximizing the use of the power transmission system." These "economic benefits" primarily accrue, however, not to consumers, but to utilities that skimp on upgrades. As late as 2005, "much of the electricity transmission system" in the United States was not computerized but relied upon "mechanical circuit breakers and controls from the 1950s."[79]

And whatever the advantages of a "free-market in electricity" may be, transparency is not one of them. Multiplying the power companies, disaggregating the ownership of generation, transmission, and retailing, and increasing traffic on high-tension lines all have made the grid more difficult to study. Scaling up the electrical system to create large grids presents unanticipated problems, and the behavior of electricity within power pools is hard to model. With each new interconnection, the current flows out to an ever-larger number of customers whose varying demands can cause fluctuations and irregularities. These can be further aggravated by lightning strikes, human error, or technical breakdowns. The more extensive the system, the harder it is to model or predict the behavior of electricity within it. Yet simulations do predict blackouts at close to the historical rate.[80]

It seems that "big blackouts are a natural product of the power grid," and that "the culprits who often get blamed for each blackout—lax tree trimming, operators who make bad decisions—are actors in a bigger drama, their failings mere triggers for disasters that in some strange ways are predestined."[81] From this macro-level viewpoint, blackouts resemble earthquakes along an active fault line. One can invest in equipment and safety procedures, but they will still occur.[82] No one can predict precisely when an earthquake or a blackout will come, but statistics can suggest when one is due. John Doyle's calculations suggested that a major blackout ought to occur about every 35 years. His result was close to a prediction of the 2003 blackout, which came 38 years after that in 1965.

Both historical experience and theoretical study suggest that blackouts are an inescapable characteristic of the electrical system, much as periodic forest fires are part of natural ecology. Elimination of small fires increases the amount of fuel available, eventually feeding a larger and more destructive conflagration. Researchers have begun to recognize that, for electrical systems, "even when mitigation is effective and eliminates the class of disruptions for which it was designed, it can have unexpected effects." In simulations, greater controls over line outages reduce small blackouts, but they increase the frequency of large blackouts. Strangely, "the overall risk may be worse than is the case with no mitigation."[83] Research also reached a counter-intuitive result about the effect of increasing generation capacity. One might expect that producing more power would reduce the chance of outages. However, when "we increase the generator margin, the character of the blackouts changes. When the generator margin is small the blackouts are small with mostly no line outages. However, at a high generation margin, they became considerably less frequent but have a large size with many line outages."[84]

The public and the politicians want to blame a culprit for each blackout, because they imagine a Newtonian world of cause and effect, of crime and punishment. Occasionally, flagrantly guilty parties such as Enron deserve to sit in the dock. But without malicious intent or incompetence a blackout may still occur. Indeed, "apparently sensible efforts to reduce the risk of smaller blackouts can sometimes increase the risk of large blackouts."[85] The electrical grid, like any other machine, will break down. The industry can prevent some failures by installing better monitoring and control devices. But as millions of kilowatts surge out from dams, nuclear plants, coal-fired powerhouses, solar arrays, biomass burners, and windmills through a vast US transmission system that could circle the earth seven times, why should anyone believe it will always work?

Indeed, terrorists may try to make it fail.

In 1996 a terrorist cell prepared to attack the electrical system of London and southeast Britain.[1] The six men between the ages of 33 and 44 were hardly novices. Among them were several senior commanders of the Irish Republican Army. They had diagrams of six power substations in a ring around London. They had detonators and fuses, and by mid July they had constructed, but not armed, 37 bombs. They planned to attack during that summer, presumably on a hot day when the system was straining to meet high demand. They would destroy transformers and other equipment, setting off a cascade of transmission problems. One conspirator was a former US Marine with training in how to plant bombs and destroy power plants behind enemy lines. The British Secret Service watched the plot unfold until July 15, 1996, when they arrested the conspirators in three simultaneous raids. Police found the bomb detonators being charged in a basement, but they never located the estimated 100 kilograms of explosives required. After a three-month trial, the six conspirators were sentenced to 35 years in prison. The plot might well have succeeded. In the same year, the IRA set off a powerful bomb in Manchester that devastated the central shopping district. It was hardly the first group to conceive such plans. In 1979 the Puerto Rican Fuerzas Armadas de Liberacion National (FALN) threatened to attack the Indian Point nuclear plant, north of New York City, and in 1981 the Macheteros bombed the Air National Guard Base at Isla Verde and later knocked out power to 20,000 customers in San Juan.[2]

When a blackout strikes today, time stops, plans fall apart, and fears fill the sudden void. A generation ago, the immediate question when the lights went out was whether a fuse had blown or lightning had struck. But since September 11, 2001, people caught in any blackout wonder if they are under attack. During the first minutes of the 2003 blackout, a reporter in Times Square saw "the bewildered, disoriented throngs, frightened by thoughts of terror" who "were trying to get their bearings in an environment that had been transformed in an instant."[3] The Federal Bureau of Investigation shared these worries,[4] because the previous year it had concluded that terrorists were studying weaknesses in power grids.[5] It put eight field offices to work, but found nothing to suggest an attack. Likewise, the FBI's Cyber Division found "no indication" that "the blackout was the result of a malicious computer intrusion or any sort of computer worm or virus."[6] Yet the FBI seriously considered those possibilities.

Shortly before the blackout, James Robbins, an editor at *National Review*, participated in an imaginary terrorist exercise. They "devised a campaign plan comprised of a series of attacks designed to spread the maximum amount of chaos at the minimum cost." Their "opening salvo was to mount a series of attacks on the utility lines near Buffalo, New York,"[7] provoking a cascading blackout on a hot summer day during rush hour. They imagined guerilla actions during the ensuing confusion. The electrical engineering publication *IEEE Spectrum* outlined a similar scenario. It suggested that after provoking a blackout, a militarist group might target crowds trying to get home on foot:

As thousands of homeward-bound pedestrians surge onto the Brooklyn Bridge, a young man suddenly tosses a grenade into the heart of the crowd. Pandemonium breaks out, as people attempt to escape the jammed walkway. In the horrifying stampede that ensues, hundreds are trampled and a few dozen jump or fall to the river below. And it isn't just the Brooklyn Bridge. The ferries and all five of the major bridges connecting Manhattan to its neighboring boroughs are attacked by grenade- and assault-rifle–toting terrorists. The eventual toll far exceeds 9/11's.[8]

The first moments of the August 2003 blackout seemed to mimic such scenarios to a disquieting degree. The breakdown originated in Ohio, in a part of the grid remote from the large cities most affected. It occurred on a hot afternoon, during rush hour, sweeping through in less than 5 minutes, beginning at 4:10. Transmission system operators in Albany, who oversee the New York State system, were astonished by an unexpected 800-megawatt surge toward the west, which then suddenly reversed direction and hurtled back toward New York. To protect themselves, they shut down generating plants and transmission lines across the Northeast. In the Albany control room, a giant board that represents the New York State system normally shows only a few red areas, indicating transmission lines under repair. But by 4:14 the whole board was flashing.[9] The next morning, al-Qaeda falsely claimed it had caused the blackout, and probably convinced some sympathizers that it had. Certainly, when the lights first went out, many Americans, including utility administrators, were prepared to believe terrorists were responsible. "Immediately after the lights flickered and went out," the *Richmond Times Dispatch* noted in an editorial,

Figure 6.1
People walk home across the Brooklyn Bridge, August 14, 2003. *Newsday Magazine.*

Figure 6.2
The Manhattan skyline at dusk during the blackout of August 14, 2003.
Newsday Magazine.

"suspicions turned to terror as a possible cause."[10] The magnitude of the recent 9/11 attack made terrorism the first interpretive framework that people adopted. Fortunately, radio news stations had standby power and informed listeners that the blackout was only what it appeared to be. There was no panic, riot, or lawlessness. As Robbins hopefully concluded, "If there is a lesson al Qaeda can draw from this event, it is that they will have to do something a lot more spectacular than even this massive power outage to get the country's attention."[11] But this assessment was too upbeat. One website asked: "What if our enemies wanted to knock power out to enhance a biological or physical attack? After all, such an outage would surely thwart emergency responders and health-care providers. It's a scenario with disastrous implications."[12] Representative Christopher Cox (Republican, California), chairman of the House Homeland Security Committee, emphasized the secondary consequences of a power loss to the country's public health, food, and water supply: "The denial of electrical service for an extended period of time causes a dangerous ripple effect of death and destruction across virtually all our nation's civic and economic sectors."[13]

Ordinary citizens took the 2003 event in their stride once they knew that the cause was an all-too-familiar combination of computer error, inadequate communication between utilities, failure to trim back the trees beneath power lines, and unforeseen interactions between the grid's far-flung components. There was a silver lining of sorts. Law enforcement officials and utilities realized that the blackout, though inconvenient and expensive, was a giant "fire drill" that tested their readiness for an actual attack. Police and fire departments went on high alert, and institutions found out whether their emergency generators or battery systems really worked. The results were mixed. The police turned out immediately, for example, but cell phones did not function, which upset almost everyone amid the general paralysis.

Most people are content to be alone when the power is on but immediately seek others when it fails. They ask for information and reassurance. As long as the electricity brings light and communication, the regular pulse of electrons is an indirect assurance that society is humming along and that a safety net of services surrounds and protects. Disrupt the flow and people immediately look for human contact. Adding the threat of terror to a power outage amplifies both fear and the desire for community. In a post-9/11 world, people immediately sought proof that the inconvenience was temporary, that the causes were, relatively speaking, benign.

Americans have long been concerned about terrorism. After the 1965 blackout, the federal government "focused increased attention on the vulnerability of power systems to disturbance and damage from acts of sabotage."[14] Indeed, "the Provost Marshall General of the United States Army conducts an annual survey of several hundred facilities in the United States which supply electric energy to important defense production areas."[15] But by far the most serious threat was terrorism directed at nuclear plants or using stolen nuclear fuel. Several attempted attacks on nuclear plants occurred between 1969 and 1971, when "explosives were discovered at a research reactor at the Illinois Institute of Technology and at the Point Beach commercial nuclear power plant, and bombs detonated in the Stanford University Linear Accelerator caused substantial damage."[16] Nevertheless, the isolated location and solid construction of nuclear facilities probably would limit civilian casualties from a successful attack.

Far more dangerous were "dirty bombs" that could spread deadly plutonium particles over a wide area.[17] The Atomic Energy Commission concluded as early as 1974 that "the potential harm to the public from the explosion of an illicitly made nuclear weapon is greater than that from any plausible power plant accident, including one which involves a core meltdown."[18] The greatest danger from a terrorist attack was from nuclear contamination, not a blackout, and for decades the US government closely monitored the location and use of every ounce of plutonium. In a 1996 attempt to understand dangers to the infrastructure more comprehensively, President Bill

Clinton established a Commission on Critical Infrastructure Protection. In addition to the electrical system, it examined telecommunications, gas and oil storage and transport, water supplies, and emergency systems, all of which require stable electricity. The commission's report led to creation of agencies to deal with infrastructural threats, which the Department of Homeland Security later absorbed.

However, not enough was actually done to improve the electrical grid between 1996 and 2003. In frustration, well before the failure of August 2003, the North American Electric Reliability Council, an industry group, warned that the grid was not being upgraded or expanded, and concluded that "the question is not whether, but when the next major failure of the grid will occur."[19] Washington officials only paid attention after the 2003 blackout. A few days later, the Secretary of Energy declared that he was confident that there would not be another such blackout if $50 billion were invested in the transmission system, with the costs passed on to consumers. However, Congress had not yet passed an energy bill to reconstruct the grid; even if it had, rebuilding would have taken a decade.

The sociologist Lee Clarke noted how institutions often issue such reassurances and create detailed crisis instructions, although these preparations are largely "fantasy documents."[20] Many civil defense plans to deal with nuclear attacks are implausible, Clarke noted, including one for commandeering commercial aircraft to evacuate the entire population of New York City in three days—a comforting but impossible scenario. Many plans narrowly focus on the point of failure and lose sight of the complexity of the problem. If "the critical infrastructure is made up of those systems required to maintain life,"[21] then it extends considerably beyond police, fire departments, and repairmen. In the influenza outbreak of 1918, hospitals, morticians, and graveyards were overwhelmed with dead bodies. As corpses literally piled up, the population became demoralized, normal life ceased, and in some communities public services collapsed. People fled, seeking to escape the disease, often leaving their dead to rot. Clarke argues that those who try to plan for "worst cases" tend to define preparations in narrow, technical terms and to focus on imagined "first responders" (e.g., police and firemen). Disaster plans typically overemphasize technical repairs and maintaining public order but overlook the centrality of actual first responders, many of whom work for schools, taxi companies, bus companies, churches, mortuaries, hospitals, and cities, but some of whom may be random passersby. Plans seldom integrate a wide range of institutions into contingency plans or recognize that in practice the first to respond are people accidentally on the scene, who must improvise. In Clarke's view, "social networks rather than formal organizations" are "far more likely" to save a life or evacuate an area in time.[22]

The collapse and blackout of New Orleans after Hurricane Katrina (2005) shows how the definition of critical infrastructure extends beyond the technical system to include a wide range of people and institutions. After Katrina hit Louisiana and Mississippi, 1.8 million people were without power. Major hospitals and homes for the elderly had to be evacuated. The Superdome, a football stadium designated as a safe haven for those who could not escape the city, also lacked electricity. About 20,000 people sweltered there in a fetid atmosphere, many suffering from heat exhaustion. And as in the London Blitz or the 1977 New York blackout, some people took advantage of the evacuation to loot, seeking not only survival necessities but also luxuries. Even hospitals were attacked. Without electricity to light the streets, sound alarms, or run pumps and other equipment, the disaster worsened each day. The National Guard went in to restore order, but much of the city remained unlivable.

As with Hurricane Katrina, the official response to disaster is often slow, disorganized, and inadequate. "Worst case disasters are too unexpected and overwhelming for organizations to fold into their standard operating procedures."[23] When the levees broke and the flood came, the Federal Emergency Management Agency failed to respond to the human needs of those trapped in the city. While FEMA dithered, volunteers saved thousands.[24] Six months after the hurricane, some New Orleans neighborhoods were still without electricity and uninhabitable.

Without electricity, present-day life loses most of its critical infrastructure. Yet the US electrical system is hard to protect because of its sheer extent. A hurricane strikes the grid randomly, but the saboteur strikes a vulnerable point where severe damage can be done, transforming social spaces that sustain life into anti-landscapes that do not. Though American grids have not been attacked, it is not because they are impregnable. Perhaps Alfred Hitchcock's 1936 film *Sabotage* suggests why transmission systems are not yet a target of choice. The film begins with a blackout that darkens a whole district of central London. A saboteur has thrown sand into some powerhouse equipment, and it takes an hour to clean and restart the system. The public response to this unexpected darkness, however, is not fear but nonchalance and even considerable laughter. Like the crowds in the 1965 New York blackout three decades later, Hitchcock's Londoners take it with aplomb. Frustrated at this result, the saboteurs then decide to bomb a crowded public place. This seems more likely than a blackout to cause panic. Even in 1936 it was clear that short-lived power outages cause few or no deaths and little destruction, whereas a bomb causes both and is terrifying. A blackout, far from being demoralizing, may strengthen the bonds of community. Accounts of the 2003 New York blackout suggest that this is what happened. A sociologist who happened to be in Brooklyn when the blackout began spent hours walking the streets, where he found people extremely helpful and far less reserved toward strangers than usual.[25]

After the rolling blackouts of the 1990s, and especially after the terrorist attacks of September 11, 2001, the public was not likely to take blackouts lightly. Rather, they improvised moments of solidarity, based on the implicit belief that the power would soon come on again. Yet the public knows that prolonged system failure will not merely paralyze a city but will also threaten them with food shortages, dehydration, and the failure of essential services, while rendering office buildings and apartments uninhabitable. Nevertheless, New Yorkers' behavior during the 2003 power failure often recalled that of 1965. Once initial fears of a possible terrorist attack were dispelled, people flocked into the streets, which took on a holiday air. The blackout became a carnival, not an apocalypse. The Russian critic Bakhtin once wrote that during a carnival "people who in life are separated by impenetrable hierarchical barriers enter into free and familiar contact."[26] Something similar occurred in New York in 2003. Even after a decade with many small power failures, a blackout could still become a moment of sociability and friendliness. As in 1965, few people could work without electricity, and all had to negotiate a city without most of its amenities. At Muldoon's Irish Pub on Third Avenue, "the loss of power meant a license to party."[27] Patrons ordered extra beer, and many ambled outside, glasses in hand. As was recalled in a blog titled "The Gothamist," at first the event "made New Yorkers wonder if there was another terrorist attack" but "then they just settled in for some street parties after finally making it home." Many brought out battery-powered audio players, sat on their front stoops, and partied into the night. Afterward, *The New Yorker* published a cartoon that suggested both the solidarity that developed among those trapped in the city that day and also how quickly that unity dissolved again. A homeless man in ragged clothing chases after a businessman, calling out "Don't you remember me? During the blackout we slept on the same sidewalk." According to one interpreter, the "August 14 event was a bit like the medieval Feast of Fools, the Yuletide holiday when in towns around Europe class distinctions were suspended, if only for a day, and masters and servants switched places, church observances were mocked, and revelry overruled solemnity."[28]

For anyone moving about, the city was "re-materialized." A visiting Brazilian architect later wrote: "Forget Virilio and Baudrillard and the virtual realities, there is no compression of time and space anymore. You are left alone with the *disvirtual* reality of space."[29] Suddenly it was not possible to mediate one's relation to the built environment, which had to be measured by the body and its ability to climb, to walk, and to adjust. "Without neon lights and electronics, space becomes what it has always been," and one "cannot hide behind a wireless phone nor dive yourself into the Internet."[30] When electrified space is decompressed, the world suddenly seems populated by unavoidable others. "Others on the stairway, Others down the street, Others on the way home. It's dark, and as a result, you start to see more and more Other people."[31] The sheer physicality of the world and its inhabitants had become bewilderingly near.

In 2003 such disorientation did not presage riots. Predictions that future blackouts would lead to unrest, persistently made during the 1980s and the 1990s, proved incorrect. Fear of terrorism partially explains why neighborhoods that erupted in looting and arson in 1977 displayed social solidarity in 2003. But just as importantly, by 2003 New York had regained prosperity. More than 200,000 people owned inexpensive apartments that had been built or renovated in areas that had experienced rioting in 1977.[32] People whose homes are appreciating are unlikely to become instant arsonists. Furthermore, New York's policing had improved dramatically, lowering the crime rate and increasing the sense of safety. Finally, the blackout came during the afternoon, when merchants were in their stores, and millions of people were in the streets walking home.

The economic impact of the 2003 blackout was much greater for commerce and industry than for households. All businesses from Detroit to Long Island were affected. At John F. Kennedy Airport, the baggage handling system stopped, and 50,000 bags could not be scanned, located, or returned to their owners, whose flights were cancelled.[33] More than 70 automobile and auto-parts plants shut down, sending 100,000 workers home. Eight oil refineries stopped production, and the main pipeline carrying Canadian oil to the United States stopped pumping. Steel plants lost batches of molten iron that had to be dumped into slag pits, and it took four days to resume production. A Republic Steel works suffered a fire because it was unable to cool down molten iron before it burned through a furnace wall. At food-processing plants, tons of meat, fruit, and vegetables rotted.[34] The average cost of a power loss was almost $60,000 per hour per business, but some corporations lost as much as a million dollars an hour.[35]

White-collar work was disrupted just as much, the losses often instantaneous and usually irretrievable. All unsaved data was lost, and many email messages never arrived. By 2003 the Internet used 8 percent of the total kilowatts generated in the United States, and at least another 5 percent was used to run individual computers. The 2003 blackout emphasized that "without electricity, there is no IT industry."[36] The microprocessors in each personal computer used 90 watts or more per hour, and there were millions of computers in the affected area. Furthermore, the heat of these microprocessors drove up demand for air conditioning.[37] The power failure showed how dependent on electricity American society had become.

The military has long understood the centrality of electricity to society. In World War II, both the Allies and the Axis Powers attacked power plants. During the 1992 siege of Sarajevo, Serb nationalists dynamited four power-transmission lines into the city, taking electricity away from 400,000 people.[38] The US military has designed weapons to incapacitate electrical networks, including the BLU-114/B "soft bomb." When used against Serbia in May 1999, it blacked out 70 percent of the country. A similar weapon had been deployed with success in the 1991 Desert Storm operation against Iraq. Detailed information about these weapons is classified, but apparently they disperse a small cloud of carbon graphite filaments and tiny wires that cause short-circuits in transformers and switches.[39]

Just as the military routinely considers how best to attack electrical systems, ordinary Americans know that power grids, nuclear power stations, and computer systems are all potential targets for terrorists. An electrical network is built to be resilient, but an attack on certain nodal points might trigger "a cascade of overload failures capable of disabling the network almost entirely."[40] Though terrorists have shown a preference for killing human beings rather than destroying infrastructure, utilities and public authorities prepare to thwart attacks and recover from them. To provide security, computer servers, alarms, scanners, cameras, metal detectors, telephones, and phone surveillance systems all require backup batteries. Many of these had been upgraded by 2003, and the emergency generators at hospitals and police stations generally worked.

Most of society does not have backup systems, however, and two Arizona State University researchers concluded that "a cascade-based attack can be much more destructive than any other strategy."[41] Moreover, some groups have disabled critical infrastructure in other countries. For example, between 1975 and 1995 the National Liberation Front of Corsica made several attacks on Corsica's electrical system. In the Philippines, the Communist New People's Army, the Moro National Liberation Front, and the Abu Sayyaf Group all attacked the electrical grid. The African National Congress did the same in South Africa during the apartheid regime, often assaulting remote installations. Each of these groups operated inside its own country. Their goal was to weaken the infrastructure while avoiding civilian casualties, in order to embarrass the government without losing popular support.[42] In contrast, terrorists who attack another country have quite different goals: to destroy prominent national symbols and to maximize human suffering, preferably in highly visible locations. This helps to explain why the US Department of State's Coordinator for Counterterrorism, J. Cofer Black, told Congress in 2003 that the American power grid was not a primary terrorist target.[43]

Yet assaults on critical infrastructure can hardly be ruled out. They have increased worldwide from 42 during the 1960s to more than 25 a year since 1990. The targets of major attacks have been oil and gas facilities (50 percent), electrical infrastructure (15 percent), office buildings (8 percent), railways (5 percent), and a wide range of miscellaneous facilities (22 percent).[44] Fifty years ago religious groups almost never made such attacks, but between 1980 and 2004 Islamist groups targeted infrastructure 84 times. In contrast, all other religious groups combined did so only five times.[45] Moreover, al-Qaeda members have specifically listed infrastructure attacks as a primary objective when attacking industrialized countries.[46] Secular terrorist groups, whose motives are primarily ethnic, ideological, or nationalist, have caused fewer casualties. The seven most active groups, including the Irish Republican Army, the Basque separatist group Euskadi Ta Askatasuna, and the Fuerzas Armadas Revolucionarios de Colombia, have never "killed more than four people in a single attack."[47] The same cannot be said for terrorists inspired by religion, who often see their cause as a holy war that justifies both material destruction and human slaughter.

A report prepared at the Lawrence Livermore National Laboratory concluded that attacks on US infrastructure were most likely from three groups. The least likely would be an attack mounted by fringe elements of a radical ecology movement such as Earth First. Edward Abbey's novel *The Monkey Wrench Gang* briefly imagines the destruction of the Glen Canyon Dam, a major hydroelectric facility on the Colorado River whose construction provoked the wrath of many environmentalists. Likewise, Ernest Callenbach's *Ecotopia* imagined that activists in the Pacific Northwest had seized atomic weapons and broken away from the United States, with a bare minimum of bloodshed, to create an egalitarian society based on sustainability. But

American eco-terrorism has been largely confined to fiction. More worrisome are right-wing militias, who might be inspired by the 1995 bombing of a federal building in Oklahoma City to attack other public structures.

The Lawrence Livermore report focused primarily on jihadist organizations. Well before 9/11, the operatives who bombed the World Trade Center in 1993 planned car-bomb attacks on the UN Building and on New York City's Lincoln and Holland tunnels, including videotaping of the tunnels and preparation of bombs.[48] Six years later, a plot to attack the Los Angeles airport was foiled when the explosives were detected and seized at a border crossing from Canada into Washington State. As these examples suggest, religious terrorists often target infrastructure, and they have caused three-fourths of all the casualties that occurred during these attacks.[49] Nevertheless, terrorists have not shown much interest in attacking the US electrical system. It appears that "almost no targets were selected purely for their function as infrastructure" and that terrorist attacks have multiple purposes, the most important being maximizing number of casualties, psychological impact, and destruction of iconic structures.[50] Destroying a remote electrical substation is not a compelling goal, though it might be one part of a larger operation.

A terrorist cell has to consider an action's overall effect, and taking out the electrical grid probably would have only a short-term impact. Utilities have rapid-response capability to deal with ice storms, forest fires, or tornados. They stockpile replacement parts and lend one another personnel to deal with emergencies. Often the loss of a transmission line does not interrupt service at all. Even taking out a substation may have a short, minimal impact.

A single ice storm in northern New York State and southern Canada in January 1998 was far more comprehensively destructive than any terrorist attack could be. It pulled down 1,000 transmission towers and 30,000 distribution poles. To duplicate that result would require a small army of terrorists armed with thousands of bombs. Yet, despite the severity of the ice storm, few people died. Utilities rapidly made repairs and restored service in all areas, however remote, within a few weeks.[51] Moreover, in that crisis the bonds of community strengthened appreciably. After five days without power, a shelter in the town of Clarkson, in upstate New York, had "evolved into a kind of place that has largely disappeared, even in small towns and villages."[52] Cut loose from their jobs and their routines, residents huddled together in what "became a public commons, a place where people met by chance and talked about whatever was on their minds." Mostly, people talked about their improvisations for light and heat. "The stories spilled out. Some people had generators, some had fireplaces or woodstoves. Those who had were sharing with those who didn't. Multigenerational families flung far across the region were coalescing at one location." This was not the usual chit-chat: "Talk was easy when everyone was bound by the same necessities." It was a powerful, collective experience, for "while everyone's experiences differed in the particulars, they were also connected and similar." As in the 1965 New York City blackout, in losing electricity people rediscovered community.

But a blackout as long as that of January 1998 is unusual. More typical was an early winter storm in November 2003. With gusting winds of more than 65 mph, and with several tornados, it briefly knocked out power for 750,000 people in Ohio. Yet the system was up and running again shortly afterward. Terrorist attacks would not pose unusual challenges to work crews already able to make such widespread repairs. The infrastructure is not a fixed and permanent entity, but is constantly being renovated, repaired, and upgraded.[53] Though it is impossible to defend every transmission line, the system as a whole is robust. Furthermore, the average household loses power for an hour or more a year, and people are irked but not terrified at the prospect. Indeed, the probability that any individual American will suffer a terrorist attack is "an order of magnitude below the annual risk of homicide" and is almost statistically insignificant when compared to the likelihood of being in an automotive accident or getting cancer.[54]

Nevertheless, in the popular imagination a blackout can have dire consequences, leading to spectacular unrest or social collapse. Joel Konvitz has called this "the myth of terrible vulnerability," noting that it depicts catastrophes that are out of control while ignoring the human ability to improvise solutions and cope with hardships.[55] With scenarios based on the direst possibilities, popular films and novels depict the electrical system as an ideal target for terrorists and revolutionaries. After the 1977 New York blackout, Arthur Hailey's novel *Overload* described a terrorist attack on California's electrical system during a summer heat wave, when the grid was already overtaxed.[56] *Overload* was among the top five *New York Times* best sellers of 1979.[57] In his 1987 novel *Patriot Games*, Tom Clancy also explored the possibility of terrorist attacks on the grid. Clancy describes an electrical engineer in Baltimore who is determined "to hurt America" by "hitting people where they lived." The character muses that "if he could turn out the lights in fifteen states at once" he would be able to weaken public confidence in the government.[58] A similar theme underlay the

2007 film *Live Free or Die Hard* (titled *Die Hard 4.0* outside North America). A disaffected former national security employee, angry because the US government has ignored his warnings about the vulnerability of the United States to a computer attack, decides to launch one himself. He gains access to federal computers, the stock market, and even the traffic lights, but his ultimate goal is to black out the eastern states. Seizing control of the hub of the transmission system, the terrorists are shutting it down region by region when the film's hero arrives, kills them all, and provokes their leader to blow up the facility. However, *Live Free or Die Hard* exaggerates the centralization of the grid. There is no single command-and-control center that can turn off all of the East's electrical systems. Rather, as the electricity consultant Jason Makansi concluded, the US transmission system in some ways is poorly integrated: "Not only is our electricity grid 'third world' in quality, it is actually weakly interconnected." However, "a weakly interconnected grid may be beneficial when it comes to security. Disconnected systems cannot all fail together."[59] Utilities, whether public or private, remain under local control. Though they are linked, each can cut itself off from the grid, and would do so in the situation that *Live Free or Die Hard* depicts. Terrorists could more easily provoke a widespread blackout by blowing up transformers or shutting down important transmission lines, causing power surges and a cascade of automatic load shedding. This would recapitulate the 2003 blackout that rippled from Ohio to Detroit, Cleveland, Toronto, and New York. Only after that expensive failure did utilities adopt improved communication systems between regions that made it possible for operators to see what was going on elsewhere on the grid. One operator put it as follows: "Think of an air traffic controller's screen. . . . The circle we can scan just got a whole lot bigger. So if there are blips [breakdowns or irregularities] further out, we see them right away." Since electricity moves more than 100,000 times faster than passenger jets, the wider perspective was essential. With new communications systems and improved software, controllers anywhere in the system could observe the grid's activity in real time. In effect, they "looked over the shoulder" of other operators, and they got information far faster than they could have gotten it by telephone.[60] Indeed, during a crisis, phone calls can be distracting, and it is often difficult to explain verbally a rapid succession of events, which can be more quickly grasped in visual representations.

"The simple reality," the energy expert Hoff Stauffer declared in 2004, "is that if a cascade like last summer's blackout starts, it's extremely difficult to stop."[61] Electrical engineers have warned that "it is impossible to secure the whole system," which includes more than 180,000 miles of transmission lines. A "determined group of terrorists could likely take out any portion of the grid they desire."[62] The most vulnerable components are high-voltage transformers, many located in substations protected only by chain-link fences. "Any one of these transformers could be knocked out of action quickly and easily with rocket-propelled grenades or improvised explosive devices."[63] Arizona State University researchers, using mathematical models of the grid, found that cascades can be expected if a single nodal point handling a large load is knocked out in a system with a "highly heterogeneous distribution of loads" such as that in North America.[64]

Not all agree that the system is vulnerable. A terrorism analyst at Rand Corporation concluded that causing a blackout would be "technically difficult" and would require considerable knowledge of the grid. But Richard Clarke, a former security official in the George W. Bush administration, warned that the grid was vulnerable to cyber attack. *Time* likewise concluded that a "skilled hacker could disable a network of several plants without ever entering a facility." He could tamper with the monitoring and control software systems, which "often lack rudimentary security, leaving technical specifications and flaws on view to potential attackers."[65] Some terrorists have computer expertise.[66] Captured al-Qaeda computers from Afghanistan had been logged on to sites dealing with utility security. Perhaps the greatest danger would be an attack by an insider, such as a disgruntled employee or someone planted in a strategic position by terrorists.[67]

Newer terrorist groups, notably al-Qaeda, have shown an affinity for using the Internet, mobile phones, encryption codes, and other new technologies to exchange information and coordinate their activities. Osama bin Laden acquired sophisticated computer equipment just as the Internet was becoming widespread, in 1996.[68] Such groups also are sensitive to the ways these technologies may be turned against them by security agencies, which can trace networks through phone company records or by monitoring websites. At the same time, they have adopted less hierarchical structures than older terrorist groups, rejecting the military chain-of-command model for loose arrangements of cells in a flat structure. They form a network that "relies less on bureaucratic fiat and more on shared values and horizontal coordination mechanisms to accomplish" their goals.[69] A flat organization with high-tech skills might attack the infrastructure. There were examples of such activity by 2000, notably by Pakistani hackers who defaced or attacked Indian military websites.[70] In 2000, expert witnesses before Congress estimated that terrorists needed 5–10 years to develop the technical capabilities to inflict major damage on the United States.[71]

Live Free or Die Hard may prove prophetic because it presented shutting down the electrical system as essentially a matter of breaking into the computer system. It reinforced the notion circulating in popular culture that a power failure is the ultimate collapse, ripping away the underpinnings of the computerized state. This was a long way from the modernist darkness of the 1930s military blackout, which signified increasing control. The terrorist blackout had become a shadowy harbinger of chaos and paralysis. However, the visions of catastrophe that pervade popular culture overemphasize vulnerability and overlook adaptation and improvisation in the face of a crisis. Indeed, in anticipation of possible infrastructure attacks, the utilities have worked with the Department of Homeland Security to prepare contingency plans. Four years before 9/11, the Department of Defense ran an exercise called "Eligible Receiver" and "found that the power grid and emergency 911 systems in the United States had weaknesses that could be exploited by an adversary using only publicly available tools on the Internet."[72] Subsequent counter-measures are not public information, but access to utility computers is more restricted. The electrical infrastructure itself also has more defenses. The focus has been less on the transmission lines, which can rapidly be restrung, than on nodal points of vulnerability, notably substations and transformers. Important sites have been "hardened" against attack, and more barriers have been erected against intruders. In addition, scientists have recommended increasing redundancy in the system. Duplication of equipment permits rapid substitution for anything destroyed. (Unfortunately, deregulated utilities have little economic incentive to pour money into such plans.) Even when such precautions are taken, however, it remains difficult to recognize a terrorist attack. As one government report concluded, "distinguishing between a routine failure and the start of a series of planned attacks is a very difficult challenge," and yet this judgement must be made, often in seconds.[73]

Security measures also continue to focus on nuclear plants, whose fuel could poison surrounding communities, and on hydroelectric dams, whose impounded waters, if released, would result in destructive flooding as well as power failures. As for cyber-attacks, utility companies' computers are in some ways more vulnerable now than before the advent of the Internet. Until the mid 1990s, most utilities had stand-alone systems that were cut off from computers elsewhere and distinctive in design. The incompatibilities between systems and their relative isolation from one another seemed undesirable, as it inhibited cooperation and coordination between generating systems, and utilities moved toward greater homogeneity, transparency, and connectivity. They built an overlay of interconnections that improved interoperability, enhancing system operators' access to crucial information in emergencies. This helped avoid impending blackouts. However, the Internet-based communications system also opened up the electrical grid to hackers. Both the Central Intelligence Agency and private consultants have found that unauthorized outsiders can break into these control systems.[74]

In view of the many avenues of potential sabotage, researchers at Carnegie Mellon University concluded that the United States can mitigate the effects of an attack on the electrical system more easily than it can prevent one.[75] Drawing on experience with rolling blackouts in California, they suggested maximum defense of a core of essential services, including traffic-control systems, hospitals, and the police. Rather than think of the grid as a fortress to be protected at every point, they call for a resilient system with multiple ways to supply power to any given area, lessening the likely duration of any outage. This suggestion challenges the electrical infrastructure's design, which for decades has been consolidated into a small number of large central stations and long transmission lines. Yet, as Lovins and Lovins pointed out in 1982, a decentralized system with a larger number of generating stations and shorter supply lines is less vulnerable.[76] *Business Week*, which generally favors deregulation, nevertheless concluded that "widespread blackouts have huge economic costs that must be measured against the savings of a unified, deregulated market," and that "a more regional, decentralized electricity system, linked by a national grid but producing most power locally, may actually be more rational, especially in an age of insecurity."[77] In 2002, the Committee on Science and Technology for Countering Terrorism reached a similar conclusion in a book-length report titled *Making the Nation Safer*. Among its recommendations was decentralization: "Electricity generation might become more decentralized, reducing the impact of the loss of key components. The use of renewable energy resources (e.g. wind and photovoltaics) would complement this trend."[78] Likewise, a Carnegie Mellon research team concluded that "a system with many small generators . . . could be made much less vulnerable."[79] Such distributed generation systems would decouple from the grid if it went down, and would be available to restart other regions during recovery.[80]

Decentralization also takes other forms, however. When individuals install small generators in their homes the decentralization is extreme, but they are not creating a distributed generation system. As an immediate solution to the threat of blackouts, some US homeowners have installed stand-alone systems behind their homes. These "whole-house generators" spin into action whenever the power fails,[81] usually because of a hurricane, an ice storm, or a brownout. For some people, for example those who rely on oxygen machines or other medical equipment to keep them alive, stand-alone generators are a necessity. At the inexpensive end of this market, traditional diesel-powered generators and portable gasoline-powered units produce 5 kilowatts. They cost less than $1,000, run for up to ten hours without refueling, and are adequate for short emergencies. But those who want more power, more security, and less noise buy gas-turbine units that generate 15 kW or more. Less polluting than either gasoline or diesel units, they require professional installation and can easily cost as much as $15,000. After each hurricane, ice storm, or summer blackout, more families buy such backup systems. Many were installed on Long Island in the 1980s after Hurricane Gloria devastated the local power system. Hundreds were sold in the St. Louis area after two week-long blackouts.[82] In Miami after the 2006 hurricane season, demand was so brisk that customers had to wait as long as 10 months for installation.[83] As thousands of individuals make this choice, the electrical system as a whole moves toward a form of decentralization that is expensive, polluting, and uncoordinated. In contrast, a system of distributed generation could focus more on renewable energy, and use efficient gas microturbines to reduce costs, enhance security, and minimize pollution.

Nevertheless, homes with autonomous electrical systems were rational responses to a US electrical grid that as late as 2005 was ill-prepared to deal with a terrorist attack. Notably, only 63 of

America's nuclear power plants even had sirens in 2005, and a mere 17 of these would have worked during a power failure. Even facilities with as many as 97 individual sirens had no backup batteries, and no plans to install them. At the Three Mile Island plant, site of the famous 1979 accident, only slightly more than one-fourth of the sirens had battery backup.[84] This lack of preparedness is all the more striking in view of the continual attacks on the energy infrastructure in Iraq. Insurgents have become adept at blowing up oil pipelines and utility lines, and in 2005 Baghdad's electricity supply was erratic, with daily blackouts in some areas.[85] The attacks were sophisticated, demonstrating a technical understanding of the power network. Although billions of dollars of American investments have rebuilt power plants, added new generators, and repaired transmission lines, the continual sabotage reduced the electricity supply to low and erratic levels.[86] Baghdad's capacity to produce electricity had been 4,500 megawatts before the American invasion of 2003, but had been reduced to only 3,500 MW in January of 2005.[87] By the end of 2006, Baghdad was under "an electrical siege," often with only two of the nine high-tension lines that led into the capital functioning to supply power.[88] Insurgents were blowing up transmission towers faster than they could be repaired, and scavengers then carted away much of the valuable wire and steel. Power shortages and blackouts had a severe impact on security, health care, business, and public confidence. In response, the Iraqi government set up 100 diesel-powered generators in Baghdad's neighborhoods, creating a decentralized patchwork of local service. Meanwhile, as a stopgap, the United States contracted with the Bechtel Corporation to install two 100-megawatt turbines. The attacks on Iraq's electrical system demonstrate both terrorists' awareness of the crucial role of electrical infrastructure and their increasing technical expertise.

The most dangerous form of blackout remains to be discussed. In 1962, an open-air nuclear test in the Pacific on Johnston Island had an unanticipated effect: a thousand miles away, in Hawaii, power grids and radio stations suddenly broke down, and some automobiles were disabled. These surprising effects were due to gamma-ray emissions that created a powerful electromagnetic pulse (EMP) that interfered with electrical equipment. This discovery prompted the Soviet Union, the United States, and later China to construct "blackout bombs"—nuclear weapons designed to maximize gamma-ray emissions. When such a bomb is exploded, gamma rays shoot off in all directions. They can disable satellites, the microchips in cars, appliances, computers, telephones, and, of course, the power grid. The explosion probably would kill few people, other than those dependent on pacemakers and other electronic devices. At least in theory, EMP damage from one bomb exploded high over Kansas could black out the entire United States. It would paralyze communications and electrical power systems. Repairs would be difficult, time consuming, and expensive, costing trillions of dollars, according to some estimates—if there were enough replacement parts available.

In 2000, the Clinton administration established a Commission to Assess the Threat to the United States from Electromagnetic Pulse Attack. The commission reported in 2004 that an EMP attack would indeed paralyze much of the US, with devastating effects. Radio, television, and telephone systems would be destroyed. The nation would not be able to harvest, store, ship, or market food. Hospitals, schools, factories, and offices would cease to function. The *Wall Street Journal* noted that even a rather "unsophisticated EMP weapon in the hands of terrorists" ("one nuclear warhead attached to a Scud missile launched from a barge off the US coast") would be able "to shut down much of the country."[89] This study, which appeared the same week as the 9/11 Commission's report, received little attention. But one member of the EMP Commission, Lowell Wood, later emphasized that nuclear weapons and the means to deploy them "are ever more present throughout the world."[90] Wood noted that "even short-range ballistic missiles are sufficient to impose such a threat, and . . . the world is . . . awash in them." Indeed, two private US citizens were able to purchase Scud missiles on the open market and have them delivered to their homes. One California homeowner had a Scud "parked on the street like an illicit camper," and the government heard about it only because neighbors complained. "EMP," Wood soberly concluded, "is one of the few ways in which the United States can die as a nation, and die in a single event, that can be readily mounted with means that are well within the purchasing capability of even a moderately wealthy individual, let alone a transnational terrorist organization."[91]

An EMP attack "turns off electric power on very large scales, potentially nationwide scales, in a fashion in which it literally cannot be turned back on. It's not a blackout, it is a stayout."[92] Whereas a conventional blackout damages only a relatively small amount of generating and transmission equipment, an EMP bomb would destroy computer chips embedded in millions of devices. "If you don't have electric power, an awful lot of other things don't function. If you don't have communications, many other things don't function. If you don't have the ability to clear financial transactions on a daily basis, the rest of the country comes tumbling down. . . . So the cross-couplings are many and they're critical."[93] A single well-placed explosion could paralyze the US and yet leave the physical landscape intact. The country would instantly become as helpless as New Orleans after Hurricane Katrina, and with communications knocked out there would be no easy way to explain to the population what had happened. There are a few reports that Iran is developing the capability to make such a bomb,[94] although Department of Defense experts and congressional consultants have argued that only Russia and China have sufficient expertise.[95]

Fortunately, vulnerability to EMP can be reduced greatly. EMP does not affect optic cables, for example. The added cost of hardening the electrical and communication infrastructure enough to survive an EMP attack is 1 percent when buying new equipment, though retrofitting costs more. Added robustness can also protect against lightning and other causes of blackouts. It can be designed into new devices, so that as consumers replace existing appliances the new ones will be resistant to EMP. EMP-resistant devices need "more robust shielding materials, especially for the cords, cables, and/or wires that connect devices to external entities such as power supplies or networks. Cables and wires act as antennas through which an EMP travels directly into a device."[96]

In an op-ed piece in the *Washington Post*, Senator Jon Kyl (R-Arizona), who chaired the Senate's Subcommittee on Terrorism, Technology, and Homeland Security, warned that al-Qaeda was capable of launching an EMP attack. He concluded: "The Sept. 11 commission report stated that our biggest failure was one of 'imagination.' No one imagined that terrorists would do what they did on Sept. 11. Today few Americans can conceive of the possibility that terrorists could bring our society to its knees by destroying everything we rely on that runs on electricity. But this time we've been warned, and we'd better be prepared to respond."[97]

The potential scale and danger of terrorist blackouts has increased from the temporary local disturbance in Hitchcock's *Sabotage* to the possibility of affecting a city, then a region, then an entire country. The blackout has also acquired additional meanings with each repetition. For some, blackouts became a postmodern form of carnival, in which people celebrated a cessation of the city's machinery. But although a blackout can be an eruption of sociability and friendliness, it can also be a harbinger of terror, crime, chaos, or conceivably the end of high-energy civilization. The public caught in an electrical blackout has an ambiguous, heterotopian experience. Many may choose to improvise a liminal moment of social solidarity. Yet, as in New Orleans after Hurricane Katrina, the familiar city teeters on the edge of becoming an anti-landscape, a broken metropolis. Perhaps it would be more accurate to say that New Orleans briefly became like urban India or South America, where "endless improvisation surrounds the distribution of scarce water, sanitation, communications, Internet, energy and transport services."[98] Such improvisation, like all repair work, leads to small improvements and upgrades, and it has the virtue of demystifying the electrical system. When citizens recover the sense that they can intervene in the construction of the system, they can think about revamping it. Instead of being passive consumers, they can become actors in the energy network. Instead of waiting for blackouts, they can organize alternatives and become less vulnerable to either terror or natural catastrophe.

Thomas Edison died in 1931, 52 years after he first displayed his incandescent electric lighting system. As a tribute, President Herbert Hoover suggested that across the United States the lights be turned out for a short time at 7 p.m. Hoover at first asked utility companies simply to impose a brief blackout, but they immediately warned against interrupting the flow of millions of kilowatts of electricity. Generating systems, then as now, could only gradually be brought on or off line, as demand rose or fell. The system had not been built to permit a sudden cancellation and an equally sudden restoration of power everywhere at once. Evidently, Hoover thought electrical grids were analogous to the wiring in a home, and imagined turning off the whole system at the flick of a switch. But power stations must step production up and step it down again in carefully orchestrated stages as the operators take generating plans on and off line. To honor Edison, it was only possible to dim highly visible lights, such as the signs in Times Square, while continuing to operate most of the system. The invention of electrical generation, wiring, and illumination was so successful that it had become integral to every part of society. As Hoover later noted in his tribute to Edison, turning off the system "would constitute a great peril to life throughout the country because of the many services dependent upon electric light in protection from fire, the operation of water supply, sanitation, elevators, operations in hospitals, and the vast number of activities which, even if halted for an instant would result in death somewhere in the country." Hoover concluded: "This demonstration of the dependence of the country on electric current for life and health is itself a monument to Mr. Edison's genius."[1]

In 1931, despite the general awareness of electricity's importance to society, the idea of suddenly cutting off or losing power scarcely had a name. Hoover had no word for it. "Outage" was used, but it had none of the metaphorical force that "blackout" would later achieve. In good part this was because loss of electricity had not meant instant darkness during the first 50 years of electrification. As late as the 1920s there were many independent electrical systems, often with no interconnections to other power providers. If one service went down, it could affect only part of a city or region. When Edison died, "blackout" was still used only in theaters.

After Edison's death, the American electrical system was woven more tightly into a distribution grid, a system of systems upon which all else relied. For each subsequent generation, a blackout stood in contrast to a different experience of light. In the 1930s, it described an intentional, artificial darkness that contrasted with the dazzle of the Great White Way and a newly electrified world. There was a touch of heroism in a military blackout, which involved many ordinary citizens in an effort to make places illegible from the air and less vulnerable to attack. After 1945, darkness itself no longer seemed natural, and a pitch-black night had become a temporary social construction. Only at this point did the word "blackout" begin to describe comprehensive power failures. By the 1950s, artificial darkness had become an unintended, ephemeral return to a pre-electrical world that was fading from memory, in contrast to the "normal" brilliance of the skyline. During the 1970s and after, however, blackouts became less a matter of darkness than of power loss. By the 1990s, "blackouts" often occurred during the daytime, particularly rolling blackouts that rotated service among consumers rather than let the whole system go down. Over 50 years, the word's meaning had changed. Rather than the planned darkness of a wartime blackout at night, it now meant an unexpected service failure, often during the day. As the grid became more integrated, blackouts flared out as eruptions of disorder amid increasing coherence.

With each passing decade, people depended more and more on electricity, until they could not live without it. At first this did not seem problematic, because generators were becoming more efficient and transmission systems more reliable. In the 1960s a power outage seemed a temporary loss of technical control, and the great Northeastern blackout of 1965 was a lark. But the 1977 New York blackout was widely perceived as inseparable from a larger loss of political and social control. Later, rolling blackouts exposed the failure of energy policy either to build sufficient supply or to moderate demand. More recently, the shadow of possible terrorist blackouts has suggested the insecurity and latent vulnerability of centralized electrical supply.

The spontaneous public response to blackouts changed too. The small-scale blackouts of the 1920s and the 1930s were simply accepted and worked around as a technological malfunction that was not particularly surprising. The military blackout evoked solidarity in the face of possible attack. In contrast, the

1965 blackout stopped the clock and forced people to live in the present. With ordinary life halted, people replaced routines with improvisation and work with play, transforming urban social space into a site for spontaneous *communitas* in a collective rite of passage. Yet a blackout did not guarantee a liminal moment of unity. It released people from the structures of electrified life, creating the possibility for various forms of collective behavior. During the 1977 New York blackout, a significant minority of the population saw the power failure not as an occasion to come together but as an opportunity for looting and arson. In the history of blackouts, in every case the public improvised in many ways. Beyond that, a blackout is not a stimulus that provokes a specific response. Rather, there are many possible reactions, including disorientation, helpfulness, displays of civic solidarity, euphoria, panic, anger, criminality, and arson. Which reaction predominates is strongly shaped by the historical context. The public perceives blackouts through the lens of current events, and each power failure expresses its moment. The dominant response to one blackout therefore will not predict behavior during the next one.

Technological vulnerability during blackouts has increased, and the public is more aware of the dangers should the system fail. Yet because the public also has become somewhat accustomed to blackouts, panic is rare when the system fails. Public information campaigns remain important to provide guidance and minimize distress during future failures. And more blackouts are likely, because upgrading the American electrical system will take decades and because the sheer complexity of the grid ensures the recurrence of "normal accidents."[2]

During the first century of electrification, the public did not have to mobilize in response to a power failure. Now it does.

Improvisation no longer is enough. Yet disaster plans may over-emphasize technical repairs and maintaining public order but overlook the first responders—typically schools, taxis, bus companies, churches, mortuaries, and random passers-by. Because "social networks rather than formal organizations" are "far more likely" to save a life or evacuate an area in time,[3] utilities and local governments should support and develop diverse networks to deal with blackouts. In addition to fire departments and the police, these networks should include businesses, schools, and voluntary organizations.

A blackout is not merely a technical event; it is also a social and cultural disruption of routines and safeguards. It has the potential to unite people in a stronger sense of community, with neighbors and strangers working together to solve problems and forging bonds that long outlive the crisis. In exceptionally favorable circumstances, such as those of 1965, a blackout can function as a liminal space and time in which class hierarchies are temporarily suspended in a euphoric moment of integration. Alternately, blackouts that are not prepared for or that are tackled with inadequate organization can tear a community apart, requiring a generation or longer to recover. In either case, the blackout is a heterotopian moment in which an alternative city suddenly emerges from the shadows. The social meaning of a blackout extends well beyond the relatively brief period in which it occurs. The response to it can either unify or weaken a community. Technological failures expose a community's degree of cohesion. They provide snapshots of its cultural condition.

Recent blackouts in much of the world provide "snapshots" of energy scarcity and environmental problems. Worldwide, growth in energy demand confronts consumers with brownouts and power outages. Yet building more capacity leads to air pollution and global warming. Caught in this bind, consumers have reacted in three quite different ways. One way to avoid blackouts is to withdraw from the grid into self-sufficiency. Some Americans did so, starting in the 1970s as an outgrowth of the environmental movement and continuing quietly afterward. By 2008 an estimated 200,000 households were "off the grid." They had adopted a variety of technologies, including better insulation, wood-burning stoves, heat pumps, solar panels, wind turbines, methane or natural gas, and occasionally water power. Sales of solar panels have increased as much as 30 percent a year. Such households carefully monitor consumption. They have a smaller "carbon footprint" than conventional homes, and they search for more efficient ways to supply heat and electricity. In effect, they have returned to stand-alone generating systems, which were common at the dawn of electrification.

The US Department of Energy has at times encouraged such decentralization and energy independence, for example by sponsoring biannual "Solar Decathlons." In 2007, twenty teams of university students designed 800-square-foot solar homes and erected them on the National Mall in Washington. Each house had to be self-sufficient, deriving all its heat, power, and light from the sun.[4] The houses were judged by a panel of engineering and architectural experts and visited by thousands of curious citizens. They showcased new technologies, such as "electrochromic windows," linked to a thermostat, that automatically darken to block the sun's heat and light or lighten to let them in. (Figure 7.1 shows an interior view of the winning entry.) The Solar Decathlons' larger goals were to raise public awareness of

Figure 7.1
Interior view of winning entry in Solar Decathlon held in Washington in 2007. The entry was from the Technical University in Darmstadt, Germany. Courtesy of US Department of Energy. Photograph by Kaye Evans-Lutterodt.

solar energy's potential, to make it cost-competitive by 2015, and to learn from the public which designs it found most appealing. Already in 2007 students constructed homes with enough surplus energy to run electric automobiles as well as to supply a house with heat, power, and light. To encourage adoption of such technologies, the Energy Policy Act of 2005 gives a $2,000 tax credit to households that install solar systems.[5] Homes with their own wind or solar power often are not "off grid," since more than 40 states have "net metering" laws that permit them to sell surplus electricity back to utility companies. This arrangement means that expensive backup battery systems can be minimized: on dark days such households can buy power from the grid, and on sunny days they can sell their surplus and spin their electric meters backward.

Another way to avoid blackouts is to increase energy efficiency. In 2008 the US economy used only half as much energy to produce one dollar of GDP as it did in 1970. Americans were getting twice as much out of their power.[6] Considerable savings came from electronic sensors and communication devices. A study covering the years 1949–2006 concluded that every kilowatt of electricity used in communication devices (e.g. by sending emails and attachments rather than letters and packages) saved between 6 and 14 kilowatt-hours elsewhere in the economy. The delivery of physical mail has also been made more energy efficient through the use of computer programs. By optimizing its delivery routes, the US Postal Service saved an estimated 3 million gallons of gasoline a year. Overall, electronic communications equipment has been used to dampen energy demand, and "for every one kilowatt of energy used . . . approximately ten kilowatts were saved."[7] Electronic devices also were used to make manufacturing processes, appliances, and automobiles more energy-efficient. Likewise, new lighting schemes for skyscrapers illuminate only the building's upper "crown," and innovative electric billboards are powered entirely by their own solar panels and wind turbines.[8] As these examples suggest, rather than building more generating plants, Americans could discard wasteful consumption habits. Sweden, Denmark, and Holland have the same standard of living as the United States and yet use only half as much energy per person.

Most houses waste electricity around the clock. In 1997, the average home in the OECD countries used 38 watts a day just to keep appliances on standby. These 386 million households were wasting 14,634 megawatts a day.[9] Redesigned appliances could save most of this power. Furthermore, by choosing the most efficient devices on the market, each household could save a further 30 percent on its energy bill.[10] Some appliances simply are unnecessary, such as clothes dryers in the arid West or air conditioning in Maine. Most homes and appliances can be designed to use less power without sacrificing comfort or performance. Likewise, compact fluorescent lights (CFLs) use far less energy than conventional incandescents.[11] In October 2007, to raise public awareness of this fact, San Francisco's Pacific Gas and Electric gave away a million CFLs. Substituted for old-fashioned incandescents, these million CFLs not only could save money on the consumer's utility bill; they could also eliminate 200,000 tons of greenhouse gases. If all of them were used, it would be equivalent to taking 60,000 homes off the grid for a year.[12] The American Public Power Association demonstrated the value of involving consumers directly in energy-saving programs. Households can learn how to reduce their consumption and change the timing of their appliance use, thereby reducing the cost of producing electricity and eliminating the need for additional power plants.[13] When people can easily measure their consumption, they usually reduce demand, and nearly every institution can save energy and money. The Automobile Club of Southern California reduced its summer consumption of electricity by 4.8 megawatts through an in-house campaign that cost only $1,000 in award money. The Bank of America reduced its annual electricity consumption by 15 megawatts through retrofits and employee conservation efforts. "Employee outreach, which cost $83,000, contributed the most savings."[14] Hospitals, factories, and many other organizations also have achieved a good return on energy-saving programs, which contribute both to profits and to morale.[15] Once implemented, economies that increase energy intensity are likely to become permanent annual savings.

New construction maximizes energy savings even more. The US Green Building Council has developed standards for energy-efficient architecture and confers awards on buildings that do especially well relative to those standards. The highest ("platinum") award does not necessarily go to a heavily insulated bunker, however. The Genzyme Center in Cambridge, Massachusetts, with its double glass walls and its heliotropic roof mirrors that focus sunlight into a twelve-story atrium (figure 7.2), uses 40 percent less energy than a conventional building of the same size. Such projects reduce demand on the electrical grid and make rolling blackouts unnecessary. Indeed, to the extent that such buildings generate solar or wind power, they become semi-autonomous. The Green Building Council's membership now includes 15,000 organizations and 70 regional chapters.[16] Since its founding in 1994, it has advocated what President Barack Obama later endorsed as the way to make the

Figure 7.2
The atrium of the Genzyme Center in Cambridge, Massachusetts. The building received a Platinum rating from the U.S. Green Building Council. Courtesy of Genzyme Corporation.

US economy more sustainable and to reduce the country's dependence on fossil fuels.

A third response to blackouts encourages "greenouts"—symbolic reductions in demand. As early as 2001, a few Americans suggested the idea of a consumer-led voluntary rolling blackout on the first day of summer, June 21. The idea was circulated on the Internet. A typical posting read: "As an alternative to George W. Bush's energy policies and lack of emphasis on efficiency, conservation, and alternative fuels, there will be a voluntary rolling blackout" that would, at least in theory, "roll across the planet."[17] This plan did not materialize, but starting in 2007 a successful movement began in Australia and the United Kingdom in which millions of consumers held short, voluntary "greenouts" during which they limited their power usage.

A greenout is dim but not dark. It resembles the brownouts necessitated by coal shortages after World War II, but it has quite different origins. A greenout displays and celebrates abstinence. It preserves the amenities electrification affords, while reducing pollution and waste. Like the blackouts of World War II, the greenout is based on civic responsibility and asks institutions and individuals to reduce lighting voluntarily. In contrast to a military blackout, the greenout has no necessary connection to the rhythms of day and night. Indeed, it may be most useful during the high-energy demand of a broiling afternoon. In a greenout, consumers reject the role of passive, high-energy victims who await rolling blackouts. Instead, they actively curb demand. For example, in 2007 Maryland's governor and one of that state's US representatives called for "one green hour" to be scheduled "from noon to 1:00 PM" on a Saturday in July.[18] It caused little social disruption, while easing the strain on electrical generators and raising awareness that consumers can easily save large amounts of electricity. Maryland was apparently inspired by a dramatic British gesture a few weeks earlier. It also emphasized that people use more electricity than they need, and showed how easily they can cut back. On the second-longest day of 2007, June 22, the Lights Out London campaign asked all residents to extinguish non-essential lights between 9 and 10 p.m. The Houses of Parliament, the Ritz Hotel, Canary Wharf, Piccadilly Circus, and Buckingham Palace were darkened. As the BBC poignantly noted, this was "the first time the lights at Piccadilly Circus had voluntarily been switched off since World War II."[19] The temporary blackout drew attention to the need to reduce emissions of carbon dioxide in order to limit climate change. London was following Sydney, Australia. Three months earlier, more than half of Sydney's homes briefly turned off their lights, and the spectacular lights of the Opera House and the Harbor Bridge were extinguished. A radio station sponsored the London event, which was supported by environmental groups such as Friends of the Earth, by Minister for the Environment Ben Bradshaw, and by Mayor Ken Livingston of London. Voluntary darkness was celebrated in newspaper photographs of the Thames River and the Houses of Parliament darkened except for the luminous dial of Big Ben. Contrasting images showed buildings such as the London Hilton first brightly lighted and then in near total darkness.[20] Participants might have saved even more energy had they turned off their radios rather than listen to Capital Radio's self-congratulation for sponsoring the event. But it did emphasize how individuals could reduce carbon dioxide pollution. Extinguishing most of London's lights for one hour reduced carbon dioxide emissions by 380 tons. Emboldened by success, the organizers immediately called for blacking out Britain as much as possible on July 7, 2007 for periods ranging from 5 minutes to the entire day.[21] This event was less successful.

Later in 2007, a Lights Out San Francisco campaign asked residents to turn off unnecessary light and power for an hour on October 20. The Golden Gate Bridge's towers went dark, and many hotels, restaurants, and other businesses participated, including Pacific Gas and Electric. But the *San Francisco Chronicle* observed that it was "not quite a blackout," and commented: "It wasn't as if all of San Francisco's skyline went dark, as it was in October 1989 after the Loma Prieta earthquake struck. But the subtle dimming of the skyline caused some whoops of excitement and then a hush from the candle-bearing assemblage of hipster environmentalists."[22] Participation increased environmental awareness. One person commented on a website: "My roommate and I were just at your event in Dolores Park and were impressed with the great showing of support for the conservation of energy. Thanks for the music, the awareness and the organization to bring the neighborhood together for a great cause."[23] Participation was less comprehensive than in either Sydney or London, however, in part because the greenout was scheduled during a World Series baseball game. An online forum received such negative comments as "Another shining example of San Francisco style over substance," "This feels like a gesture that will make a lot of people feel good about themselves," and (incorrectly) "The power surge to turn all of these lights back on at 9 may amount to more than the savings!"[24] If the event saved energy and raised awareness, it also showed that a greenout could be a tough sell even in liberal San Francisco. Based on this partial success, the organization joined an international effort in 2008, with support from Yahoo, Google, the Sierra Club, and many local businesses.

By Christmas 2007, millions of people could see this proclamation on the Web: "We are Lights Out America, a grassroots community group now organizing a nationwide energy savings event on Saturday, March 29, 2008. On this night, we invite the entire country, including businesses, city and federal government agencies, schools, and individuals to turn off all nonessential lighting to save energy." The slogan was "All across America—1 night, 1 hour, and 1 bulb on." However, a month before the targeted night the Web offered little additional information and few signs of activity, although Atlanta, Boston, Chicago, Los Angeles, Miami, New York, San Francisco, and Seattle were all participating. Publicity did not emphasize that the effort was international, nor did it mention the previous greenouts in Sydney and London.

The American context for such symbolic events was polemical. On the Fox television network, the "junk science correspondent" Steven Milloy likened environmentalists to terrorists, questioned the reality of global warming, and fulminated against "environmental zealots" and "their anti-energy jihad against essentially defenseless coal-based electricity producers and their consumers." Milloy warned that "the lights may soon go out in Washington, D.C.—and it could happen where you live, too." He granted that demand for electricity was rising faster than generation capacity, but lambasted the idea of cutting consumption. More coal-burning plants were needed, said Milloy; otherwise "the combination of unscrupulous antigrowth environmentalists and uninformed grandstanding politicians certainly will lead to lights out for America."[25]

There was less opposition in Canada. The mayor of Toronto supported the one-hour greenout, using the more universal slogan "Earth Hour." The *Toronto Star* also endorsed it: "This event is an opportunity to show how individuals acting together as a community can have a huge impact. Ultimately, we hope it gets people thinking and talking here in Toronto and in cities around the world about real solutions to what is arguably the most important issue of our time."[26] More than 150 Canadian cities participated, and Toronto was markedly darker than usual between 8 and 9 p.m.

By comparison, the United States was only partially aware of and a minor participant in the 2008 "Earth Hour." None of the presidential candidates drew attention to the event, and the television news in most cities hardly mentioned it until the following day. Few people seemed to know about it.[27] Though Toronto and Sydney tuned off their lights at 8 p.m., as did Manila and Montreal, only a few cities in the European Union or in the United States joined them. In Chicago not only the Sears Building but also many other landmarks and businesses were dark, including the golden arches of McDonalds. Yet with the notable exceptions of San Francisco and Atlanta (figure 7.3), few American or European cities were as "greened out" as Chicago. In Britain and on the Continent, the effort was often limited to darkening iconic buildings, including Brighton Pier, the Colosseum in Rome, and City Hall in Athens. In contrast, Canada and Australia reduced energy consumption by 1,000 MW or more.[28] Symptomatically, in Dublin the Environment Department, run by a Green Party Minister, darkened the Custom House, but nearby the city's financial district seemed brighter than usual.[29]

Although in 2008 the United States mostly ignored "Earth Hour," Americans remained concerned about blackouts. In February of the same year, 3 million people between South Beach and Tampa lost power after a lightning strike triggered a

Figure 7.3
Downtown Atlanta during Earth Hour, March 29, 2008. Courtesy of World Wildlife Fund.

"cascading grid collapse." This was not unusual, as "the average US electrical utility customer experiences 214 minutes of power outage each year—compared to 70 in Great Britain and just six in Japan."[30] Each summer, Americans are urged to be prepared. For example, in 2004 Governor George Pataki of New York warned of power shortages, and asked Long Islanders in particular to curb their use of electricity voluntarily. Pataki did not use the word "greenout," but he did ask people to shut off the TV, turn down the air conditioning, do their laundry late at night, and make certain they had working flashlights, reserve water, food supplies, and other essentials.[31] He was asking people to be proactive, rather than becoming passive "victims" of blackouts.

The off-grid movement, the drive for greater energy efficiency, and the first greenouts emerged about 100 years after regional electrical grids were first linked together. As the electrical system moved from novelty to necessity, the public perception of it passed through a pattern that recurs with each important new technology. That pattern can be traced in the changing perceptions of the railroad, the automobile, the assembly line, or the computer.[32] It begins with uncritical, utopian expectations of social betterment. In the 1830s the public rhapsodized over the "sublime railroad," which triumphed over space and time. It seemed to ensure growth and prosperity, while binding the country more tightly together in an economic union. The railroad, it seemed, would bring prosperity, enhance mutual understanding, and tie regions together, thereby eliminating warfare.[33] As railroads were constructed, they became a socio-economic system with great technological momentum. That system included a host of subsidiary industries that built rail cars, forged steel rails, cut down forests, and opened new farming regions. Gradually, the public realized that while railroads enabled economic growth they also facilitated economic subordination of one region to another. If they could encourage travel and the rapid exchange of ideas, during the Civil War they could also transport troops and munitions. By 1870 railroads were the most powerful businesses in the United States, and they no longer seemed utopian. Critics denounced reckless engineering that led to horrifying accidents, watered-down railroad stocks, exploitation of workers, excessive freight rates that stifled economic growth, and railroad land holdings that were kept off the market or sold at high prices, slowing westward expansion.[34] By the 1890s Americans nostalgically recalled the days of the stagecoach and covered wagons. Early in the twentieth century they hailed the automobile as a new utopian technology that would free them from the railroads' tyranny.[35]

By 2000, the centralized electrical grid had reached the last stages of a similar pattern, and Americans looked hopefully for alternative sources of power. They had all but forgotten that electricity itself once was a utopian technology. In 1879, when Edison displayed his first incandescent bulbs, newspaper reporters came from New York City by the trainload to see them, and they wrote ecstatic full-page stories.[36] Between 1890 and 1915, spectacular lighting drew crowds to expositions and to urban commercial districts.[37] But as the electrical system achieved technological momentum and people wove it into everyday life, the excitement wore off. No sooner was electricity "normal" than criticism emerged. Should this service be public or private? Centralized or decentralized? Monopolistic or competitive? Once a consensus was reached that electricity should be a centralized "natural monopoly" that usually was privately owned, the focus shifted to regulation and rate setting. Debate still flared up from time to time about whether the system should be owned by the public, but after 1920 the electrical utilities remained "natural monopolies" as they doubled and redoubled and electricity became part of the texture of life. With the important exceptions of some municipal systems and expansion of rural service as a result of New Deal programs, private, centralized monopolies enjoyed a long period of acceptance.

The more natural the electrified environment seemed, the less people thought about it. They noticed electricity only when it failed. The public also became nostalgic for older patterns of living based on the technologies electricity had replaced. Candlelight might be a throwback to pre-modern times, but it seemed romantic, suggesting a slower-paced world in which friendship, conversation, and intimacy had time to unfold. Tourists enjoyed an atmospheric return to restored water-driven grist mills or to gas-lit historic districts like those in Albuquerque and New Orleans. Likewise, blackouts provided unexpected glimpses of earlier technological conditions, and to the extent that they brought families together around candles, kerosene lamps, and firelight they could be enjoyed (when not prolonged).

Yet increased dependence on electricity in almost every facet of work, leisure, and domestic life made blackouts less and less tolerable. Furthermore, as energy prices rose, the American public criticized utilities as "natural monopolies"—particularly after the 1970s. Critics advocated energy systems that would be smaller, decentralized, based on competition, or based on the renewable energies of the sun and wind, and such ideas began to be tried out as state utility commissions became more innovative and active.[38] Just as railroad corporations became hugely unpopular in the late nineteenth century, at the close of the twentieth century electrical utilities were accused of overcharging, poor engineering, irregular service, environmental degradation, and resistance to solar and wind power.

The occasional brownouts and rolling blackouts of the 1980s and the 1990s strengthened critiques of political-economic arrangements made in 1910. By abolishing "natural" monopolies, the deregulation laws of the 1990s were expected to "fix" the electrical system. Deregulation succeeded in some states that avoided Enron-style market manipulation, but it did not address resource scarcity and global warming. An expanding, centralized electrical system, once thought utopian, began to seem problematic.

The problem is global. The atmosphere cannot infinitely absorb carbon dioxide and smoke. Power failures and pollution are becoming frequent in India and China, where noisy, polluting stand-alone generators are common. Roughly half of the world's electricity consumers must constantly improvise power and light. India suffered the biggest blackout in world history in January 2001, when more than 200 million people went without power. More typically, Indian power failures are local. During the dry season of 2002, the hydroelectric dams held insufficient water in Bangalore, center of India's IT industry and home to 6 million people. The local utility typically had a 10 percent shortfall and cut power almost every day for one to three hours. To keep operating, the Bangalore branches of Siemens, IBM, Hewlett-Packard, Infosys, and other software development companies require backup generators.[39] This problem has since worsened as India's rapid economic growth has driven electrical demand faster than it has increased generating capacity. India expects to install 40,000 megawatts of nuclear energy by 2020.[40] Yet this will be insufficient to supply the 50 percent of India's people who live without electricity and clamor for it.

Per capita, India's 1.1 billion people consume only 561 kWh a year (the US level circa 1935), and India has an installed capacity of 135,000 gigawatts. The United States, with less than one-third India's population, strains to get by with seven times that generating capacity (950,000 GW). China, with more than four times the US population, consumes just 40 percent as much electricity.[41] But as India's and China's 2.4 billion people adopt electrical appliances and air conditioning, American-style electrical consumption may become global, bringing with it blackouts, acid rain, air and water pollution, and high carbon dioxide emissions. Centralized grids, primarily powered by burning coal, are environmentally unsustainable. The ecological problems created by acid rain do not stabilize after initial damage. Rather, they intensify. Lakes die. Minerals needed by plant life leach out of the soil. Burning coal releases mercury that poisons wildlife and concentrates as it moves up the food chain, imperiling human health.[42]

The likely alternatives to coal-fired power plants will be new kinds of energy supply (solar, wind, wave, geothermal, and redesigned nuclear plants) linked into a grid that is less centralized and less prone to cascading blackouts. European countries subsidize their solar and wind power industries, whose products have improved incrementally and are now cost competitive. Europe also has the largest windmill parks, and in 2007 Germany alone generated five times as much solar energy as the United States. Much of the world's wind power comes from installations by European companies.[43] However, US investments in these alternative energies are soaring. In Hawaii solar power is already cheaper than other methods of generation, and before 2020 this will also be the case in many parts of North America. *Scientific American* devoted an extensive article to a proposed solar power network in the American Southwest that could provide 69 percent of America's electricity and 35 percent of its power needs.[44] One key to making this system work (already used successfully in Germany, Alabama, Iowa, and Texas) is converting excess solar energy into compressed air that then can be stored in caverns, in aquifers, or in other underground sites. The compressed air can be released to drive wind turbines for up to 36 hours should direct solar energy be lacking.[45] In 2004 the largest solar installation in the world, covering 30 acres, went into operation in Mühlhausen, Germany. By 2007, Germany had built fifteen of the world's twenty largest solar arrays, and its solar equipment industry employed 40,000 people.[46] Global production of photovoltaic cells grew by 51 percent in 2007, with 2.9 gigawatts of newly installed capacity.[47]

As such examples suggest, moving from a centralized system with periodic cascading blackouts to a decentralized one is less a technological problem than a political and economic one. There are many ways both to reduce electricity demand and to build more sustainable generating systems. The problem is not a lack of means but a lack of political will. Former Vice-President Al Gore underscored this fact in July 2008 when he asked Americans to transform their system of electricity generation: "I challenge our nation to commit to producing 100 percent of our electricity from renewable energy and truly clean, carbon-free sources within 10 years." Presidential candidate Barack Obama concurred: "I strongly agree with Vice President Gore that we cannot drill our way to energy independence but must fast-track investments in renewable sources of energy like solar power, wind power, and advanced biofuels."[48] Yet such a transition is not easily made. The existing energy regime has tremendous technological momentum, including the trillions of dollars invested in gas- and coal-fired central stations, the millions of jobs within the current energy system, and the ingrained habits of those who build, maintain, and use it. Previous energy regimes were not replaced in only a decade. It took more than half a century for steam engines to supplant water power.[49] To glimpse the difficulties, consider that in California, which adopted ambitious renewable energy targets in 2002 immediately after the rolling blackout crisis, more than 90 percent of new energy generation is based on fossil fuels. Nationwide, states have set long-term goals of 20–33 percent for renewable energy, but in 2009 most were missing their targets.[50]

Overall, demand for energy is rising, global warming continues, and blackouts remain common. Centralization may have seemed a sensible approach to electricity supply from 1910 until the end of the twentieth century, but decentralization and greater local control may be the sensible approach in 2010. Privatization was effective when growth was the goal, but can private enterprise reduce demand to achieve sustainability? Energy no longer has to be marketed; it now has to be de-marketed. In 1920, when most homes were still not electrified and there were only 150 home refrigerators in New York City, local utilities built and equipped all-electric houses in hundreds of American towns, where millions of ordinary families could whet their enthusiasm for a living electrically.[51] Similar demonstration homes are now needed to show that energy efficiency is practical and economical, but who is going to build them?

Reducing energy use is also a matter of aesthetics. Voluntary greenouts could become attractive to consumers if they seemed not only an environmental necessity but also a relaxing and attractive prospect. In the pre-electric world, human beings cultivated an aesthetics that did not privilege sight to such a degree and gave more weight to the other senses. Mankind evolved before the electric light, after all, and long gave greater emphasis to taste, sound, touch, and smell. When visibility is reduced, other senses come more into play. In the midst of unanticipated blackouts, some have been struck by the visual transformation of the city and found its quiet soothing and its muted vistas beautiful. A few European restaurants have successfully experimented with serving patrons who eat in the dark. Because they cannot see what they are eating, guests focus more on the food's sound, smell, texture, and taste. When a city is darkened by an electrical blackout or greenout, it sounds different. Motors, fans, televisions, elevators, subways, and other machines all stop, and people listen more carefully. To their surprise, New Yorkers found during the 2003 blackout that a symphony of insect sounds was suddenly audible. After decades of din, however, few could identify the various crickets, katydids, and cicadas.[52] A blackout forces a reconfiguration of the sensorium, realigning perceptions of familiar places. The greenout offers the same possibility, but not as an unwelcome surprise.

The urban landscape can be improved by reducing electric light. In particular, the stars and the planets become more visible. Most urbanites scarcely see them, because excessive artificial light reflects into the night sky. Indeed, as figure 7.4 demonstrates, so much electric light escapes upwards that in satellite night photographs the coastlines and major cities are readily identifiable. The Dark Sky Association suggests ways to shield lights on streets and in parking lots so that they illuminate the ground, not the sky.[53] These concerns originated during the 1960s, when poorly designed lighting interfered with the work of astronomers at the Kitt Peak National Observatory (56 miles southwest of Tucson). Civic and educational leaders solved this problem by retrofitting streetlights and passing zoning laws that reduced light pollution. Astronomers were able to continue their observations, the city became less garish, taxpayers saved money on electricity, and Tucson contributed a little less to global warming.

Not only can electric lighting be reduced; in some situations it can be replaced. The German art historian Wilhelm Hausenstein rediscovered the attractions of non-electric lighting during World War II, when for a time he was forced to use candles:

Figure 7.4
The coastlines and major cities of the United States, when seen from space at night, are delineated by electrical lighting. This 2003 image was compiled from many satellite photographs. Courtesy of National Aeronautics and Space Administration.

Lately, the electric lights have failed. We then have to rely upon the few candles that we have saved up. Having reached a point where the few good things left assume a heightened intensity, we become aware that all objects perceived in the "weaker" light of a candle acquire an altogether different dimension. Their relief seems to be higher and deeper at the same time. It feels as if the objects have regained their original corporality, something electric lighting took away from them. Superficially, electric light makes objects appear clearer and more distinct. In reality, it overwhelms and flattens them. Its glare literally eats up their bodies, their outlines, their substance. The flame of the candle, in contrast, redeems all that was lost. By restoring the shadows to objects, it restores, so to speak, their dignity and autonomy. And its brightness is just adequate to see what they really are, their poetry included.[54]

Candlelight need not be confined to emergencies. It is still common in Scandinavia and in Germany to return to candlelight when, as the cultural historian Wolfgang Schivelbusch puts it, "there is a desire to 'turn off' our managed lives for a few hours."[55]

Lowering electricity use need not impose hardships or reduce living standards, and it could confer many advantages. Low-energy appliances and homes save money. Electric cars reduce noise and air pollution. The stars and the planets would become visible again, and the five senses might come more into balance. At the national level, if rolling blackouts were not necessary and accidental blackouts less frequent, American businesses would not lose $50 billion or more a year as a result of power failures. Building windmills, solar arrays, and electric cars would create new jobs. Internationally, greening out reduces carbon dioxide emissions, helps to reduce global warming, and ensures sustainable development.

Something like a permanent greenout is urgently needed because parts of the world are becoming so damaged that they cannot sustain human life. Blackouts, as temporary anti-landscapes of energy consumption, are structurally inseparable from more permanent anti-landscapes of energy production, including lakes that are lifeless due to acid rain, areas that have been strip-mined, and 120 nuclear waste storage sites in the United States, some of which will remain contaminated for 250,000 years.[56] After a blackout, the lights may come on again within a few hours, but because New York, London, Mumbai, and Beijing insist on the "normality" of intensely electrified space some islands in the Pacific Ocean are being submerged under rising seas and some places are becoming so polluted that they are uninhabitable. Global warming, caused to a considerable degree by how electricity is produced, is transforming whole regions into anti-landscapes.

The choices are fundamental. Will Americans continue on a high-energy binge, and treat the electrical grid's technological momentum as inevitable and inexorable? Or will Americans learn to consume less energy? Will the future hold rolling blackouts, accompanied by resource depletion and global warming? Or will Americans help to build a future of alternative power generation and greater energy efficiency? Involuntary blackouts or voluntary greenouts? Brilliance interspersed with increasing blackouts, or less contrast and more chiaroscuro? Can Americans discard the ebullient high-energy attractions of modernity, acceleration, rapid obsolescence, and the technological sublime, and embrace a more contemplative way of life that emphasizes serenity, a measured pace, recycling, and technological elegance? Blackout, or greenout? We confront an unavoidable choice between two forms of artificial darkness.

Notes

Introduction

1. US-Canada Power System Outage Task Force, *Final Report on the August 14, 2003 Blackout in the United States and Canada: Causes and Recommendations*, April 2004, available at https://reports.energy.gov.

2. See, e.g., Federal Energy Regulatory Commission, *The Con Edison Power Failure of July 13 and 14, 1977.*

3. Harold Ross, "Lights out," *The New Yorker*, May 9, 1942, 9.

4. Hughes, *Networks of Power*; Hirsh, *Power Loss.*

5. Perrow, *Normal Accidents*; Nidcic, Dobson, Kirshen, Carreras, and Lynch, "Criticality in a cascading failure blackout model."

6. Foucault, "Des espace autres."

7. Turner, *The Ritual Process.*

8. Clarke, *Mission Improbable*; Jackson, *Discovering the Vernacular Landscape.*

Chapter 1

1. Owen, "The dark side: Making war on light," *The New Yorker*, August 20, 2007.

2. Beston, *The Outermost House*, 165–166.

3. Useful early works on the development of commercial lighting include Jehl, *Menlo Park Reminiscences*, Bright, *The Electric Lamp Industry*, and Keating, *Lamps for a Brighter America*. The functionalism of these early works is insufficient, and one should also see Jackle, *City Lights*, Marvin, *When Old Technologies Were New*, Bazerman, *The Languages of Edison's Light*, and Sharpe, *New York Nocturne*.

4. Nye, ed., *Technologies of Landscape*.

5. Nye, *Electrifying America*, 29–84.

6. Nye, *American Technological Sublime*, 143–198.

7. Kant, *Critique of Judgement* [1790].

8. McKinsey, *Niagara Falls*; Sears, *Sacred Places*.

9. Nye, *American Technological Sublime*, chapters 6 and 7.

10. Robinson, *Modern Civic Art*.

11. George F. Will, "Shock of recognition," *Newsweek*, July 25, 1977, 80.

12. See Smil, *Creating the Twentieth Century*, 33–97.

13. Nye, *American Technological Sublime*.

14. Israel, *Edison, a Life of Invention*, 191–229.

15. Nye, *Electrifying America*, 51–61.

16. Ekirch, *At Day's Close: A History of Nighttime*.

17. Nye, "Electrifying expositions, 1880–1939," in *Narratives and Spaces*.

18. Nye, *Electrifying America*, 51–53, 73–74, 149–150.

19. Thomas A. Edison (interview), "The woman of the future," *Good Housekeeping Magazine*, October 1913, 436.

20. See Moore, *How to Build a Home*, 57.

21. Hughes, *Networks of Power*, 14–17.

22. Only large technical systems can achieve technological momentum, which Hughes also has applied to analysis of nitrogen fixation systems and atomic energy. See Hughes, "Technological momentum: Hydrogenation in Germany, 1900–1933," *Past and Present* (August 1969): 106–132; Hughes, "Technological momentum," in Smith and Marx, eds., *Does Technology Drive History?*, 111.

23. During the 1890s, German, English, and American engineers adopted different frequencies for alternating current. Many frequencies had been in use, including 25, 30, 40, 50, 66 2/3, 125, and 133.3 cycles per second. The final selection was somewhat arbitrary, but there were constraints. If the alternating cycle is too slow, lights flicker and irritate the human eye. Yet slower frequencies are suited to running motors. The American compromise was 25 cycles per second for large motors and 60 for lighting and other general purposes. Germany and England chose 50 cycles per second. Standardization improved each system's efficiency, but today the cost of harmonizing variant standards would be prohibitive. See Hughes, *Networks*, 128–129.

24. Hughes, *Networks*, 465.

25. Weaver, *The Hartford Electric Light Company*, 59–60.

26. Cited in Conniff and Conniff, *The Energy People*, 126.

27. Hughes, "Technological momentum," in Smith and Marx, 108.

28. Individual machines can also achieve a short-term technological momentum, but only a few technologies achieve the lasting momentum of the railway network or the electrical system. A familiar example is the video cassette recorder. Machines of many different specifications might have become standard, and most experts agree that the competing Betamax machine delivered a higher quality image. But the VCR won out in the marketing war between the two systems, and achieved market dominance and technological momentum for a generation, until the advent of DVD.

29. Wainwright, *History of the Philadelphia Electric Company, 1881–1961*, 159–163.

30. Nye, *Electrifying America*, 259–282.

31. Ibid., 238–286.

32. The anniversary was marked by a reenactment sponsored by Henry Ford and attended by President Herbert Hoover and hundreds of other notables. See Nye, *The Invented Self*, 119–136.

33. Graham and Thrift, "Out of order: Understanding repair and maintenance," *Theory, Culture, and Society* 24 (2007), no. 1, 4.

34. This cost estimate is based on data found on pp. 41–42 and elsewhere in the following report: Kristina Hamachi LaCommare and Joseph H. Eto, "Understanding the cost of power interruptions to US electricity consumers," Report for US Department of Energy, contract DE-AC03–76SF00098, September 2004. LaCommare and Eto's estimate of the annual cost of power outages is $79 billion, but they admit that the data are too fragmentary to make a definite calculation. They conclude that the lowest possible estimate is $22 billion, the highest $135. The figure of $1 billion a week is therefore a conservative estimate.

35. Rosenberg, Israel, Nier, and Andrews, *Menlo Park: The Early Years, April 1876–December 1877, The Papers of Thomas A. Edison*, volume 3, 686.

36. Nasaw, *Going Out*, 120–134.

37. Schewe, *The Grid*, 78.

38. See Primeaux, *Direct Electric Utility Competition*; Grossman and Cole, eds., *The End of a Natural Monopoly*.

39. Edison Electric Institute, *Edison Electric Institute Bulletin*, February 1938, 86.

40. Hughes, *Networks of Power*, 296–300.

41. Nye, *Image Worlds*, 135–147.

42. Hughes, *Networks of Power*, 325–331.

43. Brinson, *Always a Challenge*, 3.

44. Rose, *Cities of Light and Heat*.

45. Hughes, *Networks of Power*, 264–284.

46. See Weil, *Blackout*, 28–33.

47. Brown, *Electricity for Rural America*; Nye, *Electrifying America*, 305–388.

48. Funigiello, "Kilowatts for defense: The New Deal and the coming of the Second World War," 604–620.

49. Jameson, ed., *The US Power Market*, 45.

50. Ibid., 46.

51. Victor Davis Hanson, "Paradise lost," *Wall Street Journal*, March 21, 2001.

52. BBC, "Blackouts threaten India's Silicon Valley," May 22, 2002 (http://news.bbc.co.uk).

53. Electric Power Research Institute, *Electricity Technology Roadmap*, 1–6.

54. Barbara Carton, "Acrobatic squirrels give new meaning to the term brownout," *Wall Street Journal*, February 4, 2003.

55. Schewe, 258.

56. Paul Davidson, "Aging grids cited in blackouts," *USA Today*, July 28, 2006; Marsha Freeman, "Electric grid is reaching the end game," *Executive Intelligence Review*, September 22, 2006 (http://www.larouchepub.com).

57. American Society of Civil Engineers, "Report Card for America's Infrastructure: Energy" (http://www.asce.org).

58. Perrow, *Normal Accidents*.

59. Star, "The ethnography of infrastructure," *American Behavioral Scientist* 43 (1999), no. 3: 382.

60. "Questions for Lawrence Ferlinghetti," *New York Times Magazine*, November 6, 2005.

61. Thoreau, *Walden*, 42.

62. Frisch, *Homo Faber*, 178.

Chapter 2

1. Corn, *The Winged Gospel*. Also see Douglas, *Terrible Honesty*.

2. Letter in *Hartford Courant*, May 23, 1938.

3. The *Oxford English Dictionary* traces the use of the term in the theater to 1913, when it was mentioned in a letter by George Bernard Shaw. It was first used in a military sense in a 1935 issue of the medical journal *The Lancet*.

4. Based on searches of the digital archives of both newspapers for the years 1900–1935.

5. Knell, *To Destroy a City*, 110.

6. On bombing in World War I, see Sherry, *The Rise of American Air Power*, 12–21.

7. "Gibraltar to be darkened tonight," *New York Times*, October 3, 1935.

8. "Istanbul douses lights in its first air raid drill," *New York Times*, December 21, 1935.

9. "Tokyo to combat 'air raids' today," *New York Times*, September 15, 1937.

10. "Traffic and business are impeded by Japan's blackout of her cities," *New York Times*, Aug 5, 1938.

11. "British line to Far East is guarded," *Hartford Courant*, April 24, 1939.

12. Van Dusen, *Connecticut*, 372–373.

13. "Mayor La Guardia's report on the city's civilian defense," *New York Times*, January 5, 1942.

14. On Douhet's influence, and on other writers who espoused similar views, see Sherry, 22–29.

15. Kennedy, *Freedom from Fear*, 603.

16. Sherry, 93–96.

17. "Lights and raiders in British air test," *New York Times*, August 12, 1939.

18. "Color and the black-out," *Textile Colorist* 62 (1940), February, 106, as reported in *Journal of Home Economics*, February 1940: 414–415.

19. "Reich gets ready for 'Black Week,'" *New York Times*, September 19, 1937.

20. Knud Nordgaard, quoted in "Pacifists buy fireworks to spoil army 'blackout,'" *New York Times*, October 29, 1938.

21. "No blackout or bombs will disturb Hartford," *Hartford Courant*, May 12, 1938.

22. "Town to 'black out' in 'air raid' tonight," *New York Times*, May 16, 1938.

23. James Piersol, "Blackout fails to shield target," *New York Times*, May 17, 1938.

24. "Air defense test to thrill 1,000,000," *New York Times*, October 10, 1938.

25. "Wonderful net," *Time*, October 24, 1938.

26. "Air raid black-out defends Carolinas," *New York Times*, October 14, 1938.

27. "Four hours 'blackout' tests England's defense," *Hartford Courant*, August 11, 1939.

28. "Lights aid raiders in British air test," *New York Times*, August 12, 1939. (See also "British blackout," *Newsweek*, August 14–21, 1939, 18–20.)

29. Ibid.

30. Ibid.

31. "London set to 'dig in,' city is grim," *Hartford Courant*, August 26, 1939.

32. Wattenberg, *Statistical History of the United States*, 820.

33. Mitchen, *International Historical Statistics, Europe 1750–2005*, 611–613.

34. Silvestri, "Gli sviluppi technologici," 237.

35. H. S. Bennion, "Note on power supply in warring countries," *Edison Electric Institute Bulletin*, October 1941, 405.

36. A. V. DeBeech, "Protection of electric public utility systems against air raids," *Edison Electric Institute Bulletin*, May 1942, 186.

37. Bennion, "Note on power supply," 405.

38. "British experience in overcoming raid damage to electricity supply: Actual damage has generally been purely superficial," *Electrical Review* (London), November 7, 1941.

39. DeBeech, "Protection of electric public utility systems," 183.

40. "Science in the news," *New York Times*, May 18, 1941.

41. E. N. Keller, "How British utility accountants met the Blitz," *Edison Electric Institute Bulletin*, May 1942, 190.

42. Some of these wartime broadcasts have been digitized and can be heard at http://history.sandiego.edu.

43. Porter, *London*, 338–340.

44. This paragraph drawn from Thomas, *An Underworld at War*.

45. Calder and Sheridan, *Speak for Yourself*, 100.

46. "Los Angeles dark 3 hours in alarm," *New York Times*, December 11, 1941.

47. S. G. Hibben, "Light and darkness in national defense," *Edison Electric Institute Bulletin*, July 1942, 261.

48. Apparently first recorded long after World War II, by the Dropkick Murphys, with permission from the Woody Guthrie foundation. The lyrics are available at http://www.woodyguthrie.org.

49. Kenney, *Minnesota Goes to War*, 44.

50. Hartzell, *The Empire State at War*, 90–105.

51. "Lighting work rushed," *New York Times*, December 14, 1941.

52. Sydell Rappaport, letter to the editor, *Washington Post*, October 18, 2001.

53. Tuttle, "Pearl Harbor and America's homefront children: First fears, blackouts, air raid drills, and nightmares." See also Tuttle, *Daddy's Gone to War*.

54. Alexander Maxwell, "More about blackouts: Dissimilarities of British and American needs," *Edison Electric Institute Bulletin*, March, 1942, 78. Also see Davis M. Debard, "What can American utilities learn from British war experience?" *Public Utilities Fortnightly* 29 (1942), no. 9: 531–539.

55. "Mayor says all must be put in shape for instant darkening in event of emergency," *New York Times*, March 2, 1942.

56. "New raid test turns city 'brownout' black," *New York Times*, November 23, 1943.

57. Ponting, *1940*, 162.

58. Friedrich, *The Fire*, 9, 71, 83.

59. "Australia adopts a 'brownout,'" *New York Times*, December 22, 1941.

60. I. Willis Russell, "Among the new words," *American Speech* 20 (1945), no. 2: 143.

61. "Yule trees to be lighted until 10 PM in city parks," *New York Times*, December 8, 1943.

62. "Rules for the brownout," *New York Times*, October 30, 1943. See also "Code on brownout puts sharp limit on city's lighting," *New York Times*, October 30, 1943.

63. "WPB aide assails brownout cheats," *New York Times* December 11, 1943.

64. "Blackout rules proposed," *New York Times*, January 12, 1945.

65. "No baseball brownout," *New York Times*, March 8, 1945.

66. A. F. Dickerson, "Blackouts and dimouts in the United States," *American City* 57 (1942), October: 89. See also L. A. S. Wood, "Pedestrians should be seen and not hurt," *American City* 52 (1937), November, 135; "Blackouts: Liability of city for accidents caused by such lighting restrictions," *American City* 56 (December 1941), 107; "Are dimouts necessary?" *American City* 58 (1943), April, 93.

67. Words and music: Bennie Benjamin, Sol Marcus, and Eddie Seiler.

68. "Lights fail to go up in London, darkness remains," *New York Times*, April 29, 1945.

69. Ibid.

70. "White Way and Liberty's torch leap aglow to signal victory," *New York Times*, May 9, 1945.

71. Norman Rochester, WW2 People's War, an online archive of wartime memories contributed by members of the public and gathered by the BBC, ID: A4363300, contributed July 5, 2005 (bbc.co.uk).

72. Calden, *The People's War*, 567.

73. Le Corbusier, *The Radiant City*, 178.

74. "Union blackout dims White Way," *New York Times*, August 6, 1941. See also "Union Orders Times Square Blackout," *New York Times*, August 5, 1941.

75. Louis Starks, "Kansas City is lit as tie-up goes on," *New York Times*, September 18, 1941. See also "Blackout in Kansas City," *Time* 38 (Sept 29, 1941), 16.

76. "Jersey strike deferred," *New York Times*, November 28, 1945.

77. Cited in Weber, *Don't Call Me Boss*, 219.

78. Cited in ibid., 223–224. See also Lorant, *Pittsburgh*, 491–492.

79. Goldberg, *Technological Change and Productivity in the Bituminous Coal Industry, 1920–1960*, 103; Wattenberg, *Statistical History of the United States*, 820.

80. McCraw, "Triumph and irony—the TVA," 1372–1380.

81. "Mayor calls 'brown-out' in New York," *Washington Post*, Feb 7, 1946.

82. "Soft coal in US at 'acute' level as strike goes on," *New York Times*, April 29, 1946, 1.

83. "Strike May Bring Brownout Back," *New York Times*, May 2, 1946; "Brownout described," *New York Times*, May 11, 1946.

84. "Anthracite Walkout Still Due Tomorrow," *New York Times*, May 29, 1946.

85. "Indiana imposed brown-out; confusion reigns in Chicago," *New York Times*, May 3, 1946; "No coal, Chicago hit hard," *New York Times*, May 5, 1946.

86. "Fuel-Saving Order Issued by O'Dwyer," *New York Times*, November 25; "Brownout described," *New York Times*, November 26; "O'Dwyer suspends brownout," *New York Times*, December 8.

87. "Crunch—and crisis," *Time*, May 6, 1946, 9.

88. "Historians attack Capitol blackout," *New York Times*, November 23, 1946.

89. "Topics of the Times," *New York Times*, December 9, 1946.

90. "Blackout in Tokyo," *New York Times*, October 21, 1946.

91. "Report from Osaka," *New York Times*, October 21, 1946.

92. "Japanese resent electric strikes," *New York Times*, October 24, 1946.

93. "Japan averts blackout," *New York Times*, December 2, 1946.

94. Harold Callender, "France to reduce electricity," *New York Times*, November 19, 1946.

95. "Britain's industry virtually at halt, long crisis likely," *New York Times*, February 11, 1947; "All Britain in grip of new power cut," *New York Times*, February 13, 1947.

96. Stamp, "Britain's coal crisis," 179–193.

97. Goldberg, *Technological Change and Productivity*, 103.

Chapter 3

1. "Connecticut areas in power blackout," *New York Times*, May 12, 1947. Ten years later, the word "power" was no longer needed in a headline—see, e.g., "Blackout," *Time*, September 2, 1957, 55.

2. Based on a survey of the complete digitized edition of the *New York Times* for the years 1946–1965.

3. Mumford, *The Myth of the Machine*, 409.

4. See Lurkis, *The Power Brink*, 51–55.

5. Peter Khiss, "Power fails, tying up Midtown 41/2 hours," *New York Times*, June 14, 1961.

6. Comprehensive surveys of complete runs of the *London Times* and the *New York Times* show that "blackout" was scarcely in use at all before 1930, and that it referred to darkening the lights in theaters.

7. "Lights out, city paralyzed for hours north of 59th Street," *New York Times*, January 16, 1936.

8. Ibid.

9. "Newark dark for hours as fire cuts off power," *New York Times*, December 29, 1936.

10. On how electricity was used to change the factory, see Nye, *Electrifying America*, 185–237.

11. Nye, *Consuming Power*, 157–167.

12. Cohen, *A Consumer's Republic*.

13. Nye, *Electrifying America*, 238–286.

14. *Official Guide: New York World's Fair, 1964/1965*, 202–204, 220–222.

15. de la Pena, *The Body Electric*.

16. See Nye, *Consuming Power*, 89–90; *Electrifying America*, 155–156.

17. White, *The Science of Culture*, 368.

18. Ibid., 366. See also Cottrell, *Energy and Society*. On the widespread perception that growth in energy consumption was essential to economic progress, see Hirsh, *Technology and Transformation in the US Power Industry*, 47–81.

19. For further discussion, see Marling, *As Seen on TV*, 275–278.

20. Hilgartner, Bell, and O'Connor, *Nukespeak*.

21. Nye, "Electrifying Expositions," in *Narratives and Spaces*, 113–128; *Official Guide: New York World's Fair, 1964/1965*, 220–222, 90–92, 102, 129.

22. *Century 21: A Glimpse at the 1962 Seattle World's Fair*, available at http://www.geocities.com.

23. Cowan, *More Work for Mother*, 104–105 and passim; Matthai, *An Economic History of Women in America*, 160–162.

24. Zachmann, "A socialist consumption junction."

25. Jakle, *City Lights*, 246–254.

26. See Stern, *The New Let There Be Neon* (revision of Stern's *Let There Be Neon*).

27. http://news.bbc.co.uk/go/pr/fr/-/2/hi/americas/3152451.stm.

28. Turner, *The Ritual Process*, 95.

29. Turner, *Dramas, Fields and Metaphors*, 39.

30. Turner, *The Ritual Process*, 95–97.

31. Federal Power Commission, *Northeast Power Failure*, 5–9. See also Friedlander, "What went wrong VIII," 84–85.

32. Martin Gansberg, "Power blackout affects nine states in Northeast and three Canadian provinces," *New York Times*, November 10, 1965; Edward C. Burns, "The men at Con Ed tell their story," *New York Times*, November 13, 1965.

33. A trove of documents on the 1965 blackout has been assembled, under a grant from the Sloan Foundation, at http://blackout.gmu.edu.

34. "The talk of the town," *The New Yorker*, November 20, 1965, 45.

35. Federal Power Commission, *Northeast Power Failure*, 38.

36. Ibid., 43.

37. Kathryn Westcott, "New York's 'good and bad' blackouts," BBC Online, August 15, 2003.

38. National Opinion Research Center, *Public Response to the Northeastern Power Blackout*, 12–14.

39. Loudon Wainwright Jr., "A Dark Night to Remember," *Life*, November 19, 1965, 35.

40. National Opinion Research Center, *Public Response to the Northeastern Power Blackout*, 11.

41. Federal Power Commission, *Northeast Power Failure*, 37.

42. *The New Yorker*, November 20, 1965, 43.

43. National Opinion Research Center, *Public Response to the Northeastern Power Blackout*, 55.

44. Wainwright, "A dark night to remember," *Life*, November 19, 1965, 35.

45. Turner, "Frame, flow and reflection, ritual and drama as public liminality," 465.

46. McCandlish Phillips, "Blackout vignettes," *New York Times*, November 11, 1965.

47. "The Talk of the Town," *The New Yorker*, November 20, 1965, 45.

48. Ibid., 46.

49. Phillips, "Blackout vignettes."

50. Anecdote from Billy Alban, recounted in 2007 at the mere mention of the 1965 blackout.

51. Phillips, "Blackout vignettes."

52. The poster can be seen at http://www.dorisday.net.

53. Udry, "The effect of the Great Blackout of 1965 on births in New York City."

54. For my analysis of this event, see Nye, *American Technological Sublime*, 272–280.

55. *New York Times*, July 4, 1986.

56. Foucault, "Des espace autres."

57. Ibid. (fourth principle).

58. See Nye, *Electrifying America*, chapter 2; Nye, "Electrifying expositions, 1880–1939," 113–128.

59. "The electric utility exhibits at the New York World's Fair," *Edison Electric Institute Bulletin* 7 (1939), no. 3: 81–87.

60. Boyd, "The first 100 days of the electric utilities exhibits at the New York World's Fair."

61. The GE slogan is widely referenced. *The Nation* cited it as an example of progressive patriotism.

62. See Nye, *Electrifying America*, 106–112, 148–150, 161–163, 314, 356–360, 367–379.

63. National Opinion Research Center. *Public Response to the Northeastern Power Blackout*, 11.

64. "A lesson of the blackout," *New York Times*, November 14, 1965.

65. Turner, *Man-Made Disasters*.

66. Gene Smith, "Utilities agree on a prediction: Statewide failures can recur," *New York Times*, November 11, 1965.

Chapter 4

1. Tom Topousis, "The brink of bankruptcy, *New York Post*, July 10, 2007.

2. Patterson, *Restless Giant*, 39–40.

3. Topousis, "The Brink of Bankruptcy." On fires in the Bronx, see Mahler, *The Bronx Is Burning*, 207–210.

4. Nagel, "Operating a major electric utility today," 986.

5. Ibid., 986, 992.

6. William Safire, "A dose of conspiracy theory," *International Herald Tribune*, November 6, 1995.

7. Boffey, "Investigators agree New York Blackout of 1977 could have been avoided," 994.

8. Wattenberg, *Statistical History of the United States*, 829.

9. Main, "A peak load of trouble for the utilities," 118.

10. Ibid., 5–6.

11. The Nixon administration underestimated the seriousness of the problem. Its Task Force on Oil Imports reported in February of 1970 that by 1980 US oil imports would reach 5 million barrels a day. In fact, this level was already exceeded only three years later, in 1973. For more on the "energy crisis," see Nye, "Path insistence."

12. See Cooper, *Air Conditioning America*.

13. Arsenault, "The end of the long hot summer," 602.

14. Ibid., 623–624.

15. Average automobile speed was 6 miles per hour slower by the end of the 1970s, but the number of cars had increased greatly. Both gasoline consumption and total number of miles driven had increased 20%.

16. Luke Howard, a British chemist, observed and precisely recorded the heat island of London in 1818 (Howard, *Climate of London Deduced from Meteorological Observations*).

17. Thompson, " 'The air-conditioning capital of the world,' " 88–103.

18. Source: http://eetd.lbl.gov. Also see Landsberg, *The Urban Climate*; Akbari et al., "Undoing summer heat island can save gigawatts of power"; Akbari et al., "Summer heat islands, urban trees, and white surfaces."

19. Blumberg, *Body Heat*.

20. "The Dark Side of Summer," *Village Voice*, May 15, 2007.

21. Smil, *Energies*, 148–149.

22. Goldberg, *Technological Change and Productivity in the Bituminous Coal Industry, 1920–1960*, 107.

23. Hirsh, *Technology and Transformation*, 98–99, 106, and 87–130 passim.

24. See also Smil, *Energies*, 147–149.

25. US Census Bureau, *Statistical Abstract of the United States*, 2008, table 896.

26. Hirsh, *Technology and Transformation*, 155.

27. Ehrlich, *The Population Bomb*, 15–67; Rocks and Runyon, *The Energy Crisis*, 176–177.

28. Meadows et al., *The Limits to Growth*.

29. Rifkin (*Entropy*, 79, 244, 248–254) argued that the world was entering a new era, and that the change was analogous to Europe's shift from wood to coal: "Every technology . . . is nothing more than a transformer

of energy from nature's storehouse. . . . The energy flows through the culture and the human system where it is used for a fleeting instant to sustain life (and the artifacts of life) in a non-equilibrium state. At the other end of the flow, the energy eventually ends up as dissipated waste, unavailable for future use." Continuing with fossil fuels would "only speed up chaos."

30. Sunoco reassured consumers by saying that it had invested $200 million of its profits in exploration and production in 1973, and doubled that figure in 1974, to fund a project to separate oil from tar sands in Alberta (*Business Week*, August 24, 1974).

31. *Business Week*, November 10, 1973.

32. Exxon also ran two-page advertisements that explained the co-operation between its nuclear division and General Electric (J. Walter Thompson Corporate Archives, Duke University, Competitive Advertisements T210, 1974, box 25, folder 9).

33. Southern California Edison, "The Energy Crisis," 1972 (Louis H. Roddis Collection, Duke University Library, box 11, folder 4–165).

34. *New York Times*, November 7, 1974.

35. *New York Times*, December 3, 1974; J. Walter Thompson Corporate Archives, Competitive Advertisements B110–B150, 1974, box 3, folder 11.

36. For a detailed account of the blackout, see Federal Energy Regulatory Commission. *The Con Edison Power Failure of July 13 and 14, 1977*, 23–38.

37. Boffey, "Investigators agree," 995.

38. Goodman, *Blackout*, 211.

39. Federal Energy Regulatory Commission, *The Con Edison Power Failure*, 1.

40. Goodman, *Blackout*, 4. Also see 214–216.

41. Ian Frazier, "A Good Explanation," *The New Yorker*, August 1, 1977, 27.

42. Boffey, "Investigators agree," 998.

43. Lawrence Van Gelder, "Power failure hits New York; thousands trapped in subway," *New York Times*, July 14, 1977.

44. Deidre Carmody, "Pathos, heroics, humor on a night to remember," *New York Times*, July 15, 1977.

45. Goodman, *Blackout*, 48.

46. Ralph Blumenthal, "No panic reported in subways among trapped passengers," *New York Times*, July 14, 1977.

47. Carey Winfrey, "For New Yorkers in the darkness, radio voices offered a comforting link to the light," *New York Times*, July 15, 1977.

48. William Sherman and Harry Stathos, "Spunk, cheer shine through," *Daily News*, July 14, 1977.

49. Lawrence Altman, "Bellevue patients resuscitated with hand-squeezed air bags," *New York Times*, July 14, 1977.

50. John T. McQuiston, "Medical center's parking lot like war zone field hospital," *New York Times*, July 15, 1977.

51. "Lack of water a high rise problem," *New York Times*, July 15, 1977.

52. "New outlook at Windows on the World," *New York Times*, July 15, 1977.

53. "Comment," *The New Yorker*, July 25, 1977, 19.

54. Robert D. McFadden, "New York's power restored slowly," *New York Times*, July 15, 1977. On fires, see Curvin and Porter, *Blackout Looting, New York City, July 13, 1977*, 24–25. For a general account, see Mahler, *The Bronx Is Burning*, 175–229.

55. Curvin and Porter, *Blackout Looting*, 4–5, 17.

56. Van Gelder, *New York Times*, July 14, 1997.

57. Curvin and Porter, *Blackout Looting*, 6–7.

58. Selwyn Raab, "Ravage continues far into the day; gunfire and bottles beset police," *New York Times*, July 15, 1977.

59. Ibid.

60. Goodman, *Blackout*, 59.

61. Quoted in Curvin and Porter, *Blackout Looting*, 43.

62. "Wednesday the thirteenth," *New York Times*, July 15, 1977.

63. Curvin and Porter, *Blackout Looting*, 21.

64. Ibid., 22.

65. Ibid., 22–23.

66. Goodman, *Blackout*, 76.

67. "A frenzy in dreadful darkness," *Daily News*, July 6, 1997.

68. Michael Daly, "My night in the darkness," *Daily News*, July 13, 2005, viewed at http://images.google.dk.

69. Herbert Gutman, *New York Times*, July 21, 1977; response letters, *New York Times*, August 3, 1977.

70. Goodman, *Blackout*, 157–160.

71. Podair, "Lights out," 272.

72. Some geographers and sociologists used the "anomie hypothesis" to explain the looting. Ernest Wohlenberg concluded that "people excluded from effective participation in the affluent society did not, and cannot be expected to, act with restraint when the enforcement of law and order is immobilized." (Wohlenberg, "The 'geography of civility' revisited," 42–43) The poor only lack "triggering incidents" to send them into the streets, for, in Wohlenberg's view, "economic hardship leads to violence when social controls are removed."

73. David Mamet, *Power Outage*, *New York Times*, August 6, 1977.

74. Jackson, *Discovering the Vernacular Landscape*, 8.

75. See Diamond, *Collapse*, 108–119.

76. Gerber, *On the Home Front*; Mazur, *A Hazardous Inquiry*.

77. Roush, Catastrophe and Control, 139–140.

78. Baldwin, "In the heart of darkness," 755.

79. Source: http://www.sing365.com.

80. Editorial, *Pittsburgh Press*, July 16, 1977.

81. Editorial, *Idaho Statesman*, July 17, 1977.

82. Editorials, *Burlington Free Press*, July 15, 1977; *St. Petersburg Times*, July 16, 1977; *Dayton Daily News*, July 15, 1977.

83. Associated Press, "Europeans explain why blackout was an American phenomenon," *New York Times*, July 15, 1977.

84. Michael Sterne, "Nation's reaction to the blackout; yawns to 'could it happen here?'" *New York Times*, July 15, 1977.

85. Ibid.

86. While campaigning for president, Jimmy Carter declared that he would not turn his back on New York City, as Presidents Ford and Nixon had during its prolonged fiscal crisis. When the blackout came, however, Carter refused to designate the city a disaster area, which would have qualified it for federal funds. Carter did not even visit the city for three months after the blackout.

Chapter 5

1. Matthew Wald, "East Coast is facing summer brownouts," *New York Times*, May 6, 1989.

2. "Seattle power back," *New York Times*, September 4, 1988; Peter Nulty, "Get ready for power brownouts," *Fortune*, June 5, 1989. See also Peter N. Spotts, "Why utilities tiptoe on the high wire of service reliability," *Christian Science Monitor*, August 14, 1996.

3. Hirsh, *Power Loss*, 123, 247, 261–269.

4. Ibid., 247–248.

5. Paterson, *Restless Giant*, 115.

6. Hirsh, *Power Loss*, 168–170.

7. Ibid., 9–70.

8. Thanks to Richard Hirsh, who pointed out the importance of efficiency gains.

9. Statistics from US Department of Energy, "Table US-1. Electricity Consumption by End Use in US Households, 2001" (http://www.eia .doe.gov).

10. Nadar, "Barriers to thinking new about energy."

11. Nadar, "The harder path."

12. Michigan Public Service Commission, *Report on the August 14th Blackout*, November, 2003, 25. Also see US Department of Energy, *Energy Information Administration Annual Energy Review, 2001*, table 8.8.

13. US Patent 4075699, issued February 21, 1978.

14. US Patent 5572438, issued January 5, 1996 to TECO Energy Management Services.

15. Smil, *Energies*, 147–149.

16. California Energy Commission, *1994 Electricity Report. Introduction and Executive Summary*, November 1995, 6.

17. Charles David Jacobson, "Electric utility restructuring in California," at www.thebhc.org.

18. Swartz and Watkins, *Power Failure*, 241.

19. Stephen Braun, "Threat of blackouts has Chicago set to lose cool," *Los Angeles Times*, July 24, 1998.

20. "Utilities outages spark outrage in Chicago," *Los Angeles Times*, August 24, 1999.

21. Scott Thurm, Robert Gavin, and Mitchel Benson, "Juice squeeze," *Wall Street Journal*, September 16, 2002.

22. Lambert, *Energy Companies and Market Reform*, 153.

23. For a summary of these illegal and unethical practices, see Weil, *Blackout*, 87–106. See also Swart and Walkins, *Power Failure*, 238–242, 248–249.

24. Jason Leopold, "Another slap on the wrist," *Counterpunch* (http://www.counterpunch.org), February 3, 2003.

25. Swart and Walkins, *Power Failure*, 242.

26. Fox, *Enron*, 208–210. See also Lambert, *Energy Companies and Market Reform*, 28–44, 170–176.

27. Lambert, *Energy Companies and Market Reform*, 117–141.

28. Ibid., 141.

29. Slocum, "Electric utility deregulation and the myth of the energy crisis," 477.

30. Lambert, *Energy Companies and Market Reform*, 2. See also Kathryn Kranhold, Bryan Lee, and Mitchel Benson, "Enron rigged power market in California, documents say," *Wall Street Journal*, May 7, 2002.

31. Schewe, *The Grid*, 181.

32. Source: http://www.eia.doe.gov.

33. Nancy Vogel and Nancy Rivera Brooks, "Rolling blackouts again hit California," *Los Angeles Times*, January 18, 2001.

34. Lively, "California colleges struggle with blackouts."

35. Evelyn Nieves, "Dark days send chill through Dairyville," *New York Times*, January 21, 2001.

36. "California IT staff prepare for more blackouts," *Computer World*, March 6, 2001.

37. For more on the California electricity crisis, see Wolak, "Lessons from the California electricity crisis."

38. Apt et al., "Electrical blackouts: A systemic problem."

39. Howe, "A year after the blackout."

40. Calviou, "Transmission: A key market enabler."

41. David Cay Johnston, "Grid limitations increase prices for electricity," *New York Times*, December 13, 2006.

42. Cited in Paul Krugman, "The road to ruin," *New York Times*, August 19, 2003. Joskow received some of his research funding from Enron. See Swartz and Watkins, *Power Failure*, 242.

43. Apt et al., "Electrical blackouts," 57.

44. Lesieutre and Eto, "Electricity transmission congestion costs," 2. Unfortunately, studies of congestion do not benefit from a standardized methodology, and it is difficult to compare them or to get a national overview of the total costs. But the order of magnitude is above $1 billion a year.

45. Matthew Bunk, "Research center at mercy of budget," *Oakland Tribune*, February 25, 2006. The budget of over $300 million in 2002 fell below that level from 2004–2006. See EPRI Annual Report, 2006 (http://my.epri.com).

46. Sovacool, *The Dirty Energy Dilemma*, 159–161.

47. "Expansion of US power grid likely to rely at first on proven, decades-old technologies," *Electric Utility Week* August 28, 2006: 3–4.

48. Wertheimer and Leeper, "Electrical wiring configurations and childhood cancer," 273; Savitz, "Case-control study of childhood cancer and exposure to 60-Hz magnetic fields," 21; David L. Chandler, "Study finds power line tie to leukemia," *Boston Globe*, November 12, 1992.

49. Howe, "A year after the blackout."

50. Johnston, "Grid limitations increase prices for electricity."

51. Cited in Weil, *Blackout*, 110.

52. US-Canada Power System Outage Task Force, *Final Report on the August 14, 2003 Blackout in the United States and Canada: Causes and Recommendations*, 2004 (https://reports.energy.gov), 17–22.

53. See, e.g., Thomas, "Managing relationships between electric power industry restructuring and grid reliability," 15.

54. North American Electric Reliability Council, "2004 long-term reliability assessment." Available at http://www.nerc.com.

55. Fairley, "The unruly grid: One year later," 5.

56. Source: http://www.nerc.com.

57. US Department of Energy, "National electric delivery technologies roadmap," 2004, 2.

58. Fama, "Lessons of the Blackout."

59. Howe, "A year after the blackout."

60. Munson, *From Edison to Enron*, 129; presentation by David Meyer, Harvard University, 2008.

61. "Stop bickering and fix the power grid," *Business Week*, September 1, 2003, 104.

62. Joyce Price, "Western blackout blamed on surge," *Washington Times*, December 15, 1994.

63. Casazza, "Blackouts," 63.

64. "Blackout in the Northeast and Midwest," Hearing before Committee on Energy and Natural Resources, US Senate, February 24, 2004 (US Government Printing Office, 2004), 22.

65. Meyer, 2008 presentation.

66. John G. Edwards, "Power shortage: Blackout procedure defended," *Las Vegas Review*, July 4, 2001.

67. "California warns of rolling blackouts," *New York Times*, July 24, 2006.

68. Peter J. Howe, "Power grid will push for energy conservation," *Boston Globe*, June 20, 2006.

69. Anne Tafton, "Electricity blackouts: A hot summer topic," MIT news office, August 8, 2006.

70. Carreras et al., "Complex dynamics of blackouts in power transmission systems," 643.

71. Makarov et al., "Blackout Prevention in the United States, Europe, and Russia."

72. Ibid., 1946–1947.

73. Ibid., 1945.

74. Jonathan Freedland, "Huge power cut leaves Americans in the dark," *Guardian*, July 4, 1996.

75. Carreras et al., "Blackout mitigation assessment in power transmission systems."

76. Nidcic et al., "Criticality in a cascading failure blackout model."

77. Ibid.

78. Ibid., 631.

79. Munson, *From Edison to Enron*, 128.

80. See Carreras et al., "Blackout mitigation assessment," 2.

81. Fairley, "The unruly power grid," 5.

82. Kirschen and Strbac, "Why investments do not prevent blackouts."

83. See Carreras et al, "Blackout mitigation assessment," 5.

84. Ibid., 8.

85. Ibid., 9.

Chapter 6

1. Warren Hoge, "Britain convicts six of plot to black out London," *New York Times*, July 3, 1997. For a map showing target locations, see John Steele, "IRA planned to black out London." *Daily Telegraph*, April 12, 1997.

2. Bruce Hoffman, *When the Lights Go Off and Never Come Back On*; Jo Thomas, "Puerto Rico terrorist group takes responsibility for blackout," *New York Times*, November 29, 1981.

3. Bob Herbert, "Still dangerously unprepared," *Biloxi Sun Herald*, August 19, 2003.

4. On the FBI, see William Jackson, "IT, power grids not primary terror targets, FBI says," *Post-Newsweek Business Information*, September 12, 2003.

5. *Newsweek*, August 18, 2003.

6. Ibid.

7. James S. Robbins, "Al Qaeda done it?" National Review Online (http://www.nationalreview.com), August 19, 2003.

8. "Nine cautionary tales," *IEEE Spectrum*, September 2006: 36–45.

9. *Newsweek*, August 18, 2003.

10. "What a riot" (editorial), *Richmond Times Dispatch*, September 24, 2003.

11. Robbins, "Al Qaeda Done It?"

12. Dennis McCafferty, "Dark lessons: Learning from the blackout of August '03," www.HSToday.us.

13. "Homeland Security panel to investigate grid's vulnerability to attack," *Washington Times*, August 18, 2003.

14. Federal Power Commission, *Prevention of Power Failure*, 65.

15. Ibid., 67.

16. Walker, "Regulating against nuclear terrorism," 111.

17. For graphic descriptions of such dangers, see McPhee, *The Curve of Binding Energy*.

18. Cited in Walker, "Regulating against nuclear terrorism," 107.

19. Cited in Herbert, "Still Dangerously Unprepared."

20. Clarke, *Mission Improbable*, 33–35.

21. Clarke, *Worst Cases*, 164.

22. Ibid., 170.

23. Ibid., 170.

24. See Brinkley, *The Great Deluge*.

25. Yuill, "Emotions after dark—A sociological impression of the 2003 blackout," *Sociological Review online* 9:3 2004.

26. Bakhtin, *Rabelais and His World*, 123.

27. Michael Slackman, "Blackout of 2003; a party at the pub," *New York Times*, August 15, 2003.

28. Schewe, *The Grid*, 19.

29. Lara, "In the Dark All You Have Left is Architecture."

30. Ibid.

31. Ibid.

32. Brent Staples, "Why once-violent neighborhoods stayed calm during the blackout," *New York Times*, August 24, 2003.

33. Moore, "High stakes, high voltage."

34. See Electricity Consumers Resource Council, "The Economic Impacts of the August 2003 Blackout," 2004, available at http://www.elcon.org.

35. LaCommare and Eto, *Understanding the Cost of Power Interruptions to US Electricity Consumers*, 38.

36. Electricity Consumers Resource Council, "The Economic Impacts of the August 2003 Blackout."

37. Moore, "High stakes, high voltage," 1.

38. John F. Burns, "Power and water lost in Sarajevo as attacks mount," *New York Times*, July 14, 1992.

39. Source: Military Analysis Network (www.fas.org).

40. Motter and Lai, "Cascade-based attacks on complex networks."

41. Ibid.

42. Ackerman et al., *Assessing Terrorist Motivations for Attacking Critical Infrastructure*, 170–171.

43. William Jackson, "IT, power grids not primary terror targets," *Post-Newsweek Business Information*, September 12, 2003.

44. Ackerman et al., *Assessing Terrorist Motivations*, 171.

45. Ibid., 107–109.

46. Ibid., 170.

47. Ibid., 110.

48. Lawrence Livermore National Laboratory, *The Jericho Option*.

49. Ibid., 19.

50. Ibid., 229.

51. Doheny-Farina, *The Grid and the Village*.

52. Ibid., 141–142.

53. See Graham and Thrift, "Out of order."

54. Smil, "The next 50 years," 209.

55. Konvitz, "Why cities don't die," 58–63.

56. Hailey, *Overload*.

57. *New York Times Sunday Book Review*, February 4, 11, and 18, 1979.

58. Clancy, *Patriot Games*, 334.

59. Makansi, *Lights Out*, 14.

60. Ian Urbina, "New York utilities recall '03 blackout with fingers crossed," *New York Times*, May 31, 2004.

61. Ibid.

62. "Nine cautionary tales."

63. Ibid.

64. Motter and Lai, "Cascade-based attacks on complex networks."

65. Johanna McGeary, "An Invitation to Terrorists," *Time*, August 25, 2003, 2.

66. Lawrence Livermore National Laboratory, *The Jericho Option*, 134–135, 237.

67. Committee on Science and Technology for Countering Terrorism, *Making the Nation Safer*, 190.

68. Schell and Dodge, *The Hacking of America*, 189.

69. Zanini and Edwards, "The Networking of Terror in the Information Age," 33.

70. Ibid., 44.

71. D. E. Denning, "Cyberterrorism: Testimony before the Special Oversight Panel on Terrorism, Committee on Armed Services," US House of Representatives, May 23, 2000.

72. Schell and Dodge, *The Hacking of America*, 189.

73. Committee on Science and Technology for Countering Terrorism, *Making the Nation Safer*, 182.

74. Lawrence Livermore National Laboratory, *The Jericho Option*, 35, 70, 157, 237.

75. Ibid., 54–56.

76. Lovins and Lovins, *Brittle Power*.

77. "Stop bickering and fix the power grid," *Business Week*, September 1, 2003, 104.

78. Committee on Science and Technology for Countering Terrorism, *Making the Nation Safer*, 179.

79. Farrell, Lave, and Morgan, "Bolstering the security of the electric power system," 55.

80. They could also be systems of "distributed generation" able to sell the heat that is a by-product of making power. Cogeneration, common in the Netherlands and in Scandinavia, is competitive with large power plants, which often simply dump heat as waste.

81. Ann Fisher, "Generators a security blanket all year long," *Columbus Dispatch*, February 16, 2007.

82. Michael Girardo, "An aid in brownouts: Portable generators," letter in *New York Times*, June 8, 1986; Makansi, *Lights Out*, 226.

83. "Whole house generators," *Miami Herald*, May 30, 2007.

84. "Most nuclear plant warning sirens lack backup power," Associated Press, May 27, 2005.

85. James Glanz, "Iraq insurgents starve capital of electricity," *New York Times*, December 19, 2006.

86. James Glanz, "Insurgents wage precise attacks on Baghdad fuel," *New York Times*, February 21, 2005.

87. James Glanz, "New election issues: Electricity and water," *New York Times*, January 26, 2005.

88. Glanz, "Iraqi insurgents starve capital of electricity."

89. "Mother of all blackouts" (editorial), *Wall Street Journal*, August 12, 2004.

90. Lowell Wood, address at Fletcher School, Tufts University, 2004.

91. Ibid.

92. Ibid.

93. Ibid.

94. Joseph Farah, "Iran plans to knock out US with 1 nuclear bomb," WorldNetDaily.com, April 25, 2005.

95. Singer, "Experts say United States is unprepared for EMP attack."

96. Patrick Chisholm, "Protect our electronics against EMP attack," *Christian Science Monitor*, December 19, 2005.

97. Jon Kyl, "Unready for this attack," *Washington Post*, April 16, 2005.

98. Graham and Thrift, "Out of Order," 11.

Chapter 7

1. Herbert Hoover, "Statement on a National Tribute to Thomas Alva Edison," October 20, 1931 (available at http://www.presidency.ucsb.edu).

2. Kirschen and Strbac, "Why investments do not prevent blackouts"; Nidic et al., "Criticality in a cascading failure blackout model."

3. Clarke, *Worst Cases*, 170.

4. Source: www.solardecathlon.org.

5. "Off the grid or on, solar and wind power gain," *USA Today*, April 12, 2006.

6. Sovacool, *The Dirty Energy Dilemma*, 80–81.

7. Laitner and Ehrhardt-Martinez, *Information and Communication Technologies*, 21, 26.

8. Ken Belson, "Efficiency's mark: A city glitters a little less," *New York Times*, November 1, 2008; Glenn Collins, "In Times Square, a company's name in (wind- and solar-powered) lights," *New York Times*, November 14, 2008. See also Steven McElroy, "The Great White Way tries to turn green," *New York Times*, November 25, 2008.

9. International Energy Agency, *Things That Go Blip in the Night*, 90.

10. Ibid., 195.

11. Whereas an incandescent bulb creates light by heating up a fila-ment, a compact fluorescent bulb contains a gas that gives off ultraviolet light when electrified. The ultraviolet light excites a phosphor coating on the glass of the bulb that then gives off light. Since the 1980s compact fluorescent bulbs have improved in quality and declined in price.

12. Pacific Gas and Electric Company Kicks Off Energy Awareness Month . . . to Give Away One Million CFLs During October, at http:// www.pge.com.

13. Cited in Hoffman and High-Pippert, "Community energy," 393. See also www.appanet.org.

14. Eggink, *Managing Energy Costs*, 191–192.

15. Ibid., 192–194.

16. Source: http://www.usgbc.org.

17. Source: http://www.snopes.com.

18. "Greenout," *Baltimore Sun*, July 25, 2007.

19. "London lights out for environment," June 22, 2007, http://news .bbc.co.uk.

20. Source: http://www.lightsoutlondon.com.

21. Source: http://www.blackoutbritain.org.uk.

22. Tyche Hendricks, "Not quite a blackout, but a good turnout for Lights Out event," *San Francisco Chronicle*, October 21, 2007.

23. Source: http://lightsoutsf.org.

24. Source: http://www.sfgate.com.

25. Steven Milloy, "Lights Out, America?" foxnews.com, February 7, 2008.

26. Catherine Porter, "Countdown to Earth Hour," *Toronto Star*, December 13, 2007.

27. I was in New England on March 28 and 29, 2008 and could find almost no one who had heard about the event. There was scant coverage in the Hartford and Springfield newspapers.

28. Peter Gorrie, "Toronto hits energy target," *Toronto Star*, March 29, 2008.

29. Shawn Pogatchnik, "Sites dim for Earth Hour," *Springfield Republican*, March 29, 2008.

30. Ted Padgett, "Florida's blackout, a warning sign?" *Time*, February 27, 2007.

31. Bruce Lambert, "How Long Island prepares for a blackout: Practice, practice, practice," *New York Times*, June 4, 2004.

32. Nye, "From utopia to 'real-topia,'" 161–172.

33. Ward, *Railroads and the Character of America*, 60–63.

34. See Nye, *America as Second Creation*, 175–204.

35. Belasco, *Americans on the Road*, 19–39.

36. Israel, *Edison*, 187–189.

37. Nye, *Electrifying America*, 33–47, 63–65.

38. See Hirsh, *Power Loss*, 89–119; Anderson, *Regulatory Politics and Electric Utilities*.

39. BBC, "Blackouts threaten India's Silicon Valley," May 22, 2002.

40. Archana Chaudhary, "Tata Power may build nuclear projects after India changes rules," http://www.bloomberg.com.

41. Source of statistics: US Energy Information Administration (http://www.eia.doe.gov).

42. Glick, "The toll from coal," 482.

43. See also Berger, *Charging Ahead*. German engineers created a "combined renewable energy power plant" that demonstrated how different forms of renewable power together can supply all of a community's power. This system linked eleven wind farms, four biogas generators, twenty photovoltaic plants, and one pumped storage facility. A centralized control system enabled these suppliers to meet a community's fluctuating energy demands. Similarly, alternative energies could eventually satisfy more than 80% of the electrical needs of the US, eliminating most air pollution in the process.

44. Zweibel, Mason, and Fthenakis, "A solar grand plan."

45. Scheer, *Energy Autonomy*, 68–69.

46. Craig Whitlock, "Cloudy Germany a powerhouse in solar energy," *Washington Post*, May 5, 2007.

47. Janet L. Sawin, "Another sunny year for solar power," http://www.worldwatch.org.

48. Source of Gore and Obama quotations: John M. Broder, "Gore urges change to dodge an energy crisis," *New York Times*, July 18, 2008.

49. Nye, *Consuming Power*, 44–51, 79–82.

50. Kate Galbraith and Matthew L. Wald, "The energy challenge: Energy goals a moving target for states," *New York Times*, December 5, 2008.

51. Nye, *Electrifying America*, 266–267.

52. Nick Paumgarten, "Department of Entomology," *The New Yorker*, September 15, 2003.

53. Internet address of International Dark Sky Association: http://www.darksky.org.

54. Hausenstein, quoted in Wolfgang Schivelbusch, "The Force," *Cabinet 21* (spring 2006).

55. Ibid.

56. Sovacool, *The Dirty Energy Dilemma*, 69–71.

Bibliography

Abbey, Edward. *The Monkey House Gang*. Harper Perennial, 2000.

Ackerman, Gary, et al. *Assessing Terrorist Motivations for Attacking Critical Infrastructure*. Weapons of Mass Destruction Terrorism Research Program, Center for Nonproliferation Studies, Monterey Institute of International Studies, 2007.

Akbari, H., J. Huang, H. Taha, and A. Rosenfeld. "Undoing summer heat island can save gigawatts of power." In *Proceedings of ACEEE 1986 Summer Study on Energy Efficiency in Buildings*, volume 2. Also Lawrence Berkeley National Laboratory Report LBL-21893.

Akbari, H., A. Rosenfeld, and H. Taha. "Summer heat islands, urban trees, and white surfaces." In *Proceedings of American Society of Heating, Refrigeration, and Air Conditioning Engineers*, Atlanta, 1990. Also Lawrence Berkeley National Laboratory Report LBL-28308.

Anderson, Douglas D. *Regulatory Politics and Electric Utilities: A Case Study in Political Economy*. Auburn House, 1981.

Apt, Jay, Lester B. Lave, Sarosh Talukdar, M. Grander Morgan, and Marija Ilic, "Electrical blackouts: A systemic problem." *Issues in Science and Technology*, summer 2004: 1–8.

Arsenault, Raymond. "The end of the long hot summer: The air conditioner and Southern culture." *Journal of Southern History* 50 (1984), no. 4: 587–628.

Bakhtin, Mikhail. *Rabelais and His World*. University of Indiana Press, 1984.

Baldwin, Peter C. "In the heart of darkness: Blackouts and the social geography of lighting in the gaslight era." *Journal of Urban History* 30 (2004), no. 5: 749–768.

Bazerman, Charles. *The Languages of Edison's Light*. MIT Press, 1999.

Belasco, Warren James. *Americans on the Road*. Johns Hopkins University Press, 1997.

Bennion, H. S. "Note on power supply in warring countries." *Edison Electric Institute Bulletin*, October 1941: 405–406.

Berger, John J. *Charging Ahead: The Business of Renewable Energy and What It Means to America*. Holt, 1997.

Beston, Henry. *The Outermost House*. Holt, 1988.

Blumberg, Mark S. *Body Heat: Temperature and Life on Earth*. Harvard University Press, 2002.

Boffey, Philip M. "Investigators agree New York blackout of 1977 could have been avoided." *Science* 201 (1978), September 15: 994.

Boyd, Gardner. "The first 100 days of the electric utilities exhibits at the New York World's Fair." *Edison Electric Institute Bulletin* 7 (1939), no. 8: 377–378.

Bright, Arthur A. *The Electric Lamp Industry: Technological Change and Economic Development from 1800 to 1947*. Macmillan, 1949.

Brinkley, Douglas. *The Great Deluge: Hurricane Katrina, New Orleans, and the Mississippi Gulf Coast*. HarperCollins, 2006.

Brinson, Carroll. *Always a Challenge: Mississippi Power and Light Company's First Sixty Years*. Oakdale, 1984.

Brown, Clayton D. *Electricity for Rural America: The Fight for the REA*. Greenwood, 1980.

Calden, Angus. *The People's War, Britain, 1939–1945*. Jonathan Cape, 1969.

Calder, Angus, and Dorothy Sheridan. *Speak for Yourself: A Mass Observation Anthology, 1937–1949*. Jonathan Cape, 1984.

California Energy Commission, 1994 Electricity Report. Introduction and Executive Summary, 1995. http: //www.energy.ca.gov.

Callenbach, Ernest. *Ecotopia*. Bantam, 1977.

Carreras, B. A., V. E. Lynch, Ian Dobson, and D. Newman. "Complex dynamics of blackouts in power transmission systems." *Chaos* 14 (2004), no. 3: 643.

Carreras, B. A., V. E. Lynch, D. E. Newman, and I. Dobson. "Blackout mitigation assessment in power transmission systems." Presented at Hawaii International Conference on System Science, January 2003. IEEE.

Casazza, John. "Blackouts: Is the risk increasing?" *Electrical World* 212 (1998), no. 4: 62–64.

Clancy, Tom. *Patriot Games*. Berkeley Books, 1987.

Clark, Lee. *Mission Improbable: Using Fantasy Documents to Tame Disaster*. University of Chicago Press, 1999.

Clarke, Lee. *Worst Cases: Terror and Catastrophe in the Popular Imagination*. University of Chicago Press, 2006.

Cohen, Lizabeth. *A Consumer's Republic: The Politics of Consumption in Postwar America*. Harvard University Press, 2003.

Committee on Science and Technology for Countering Terrorism. *Making the Nation Safer*. National Academies Press, 2002.

Conniff, James C. G., and Richard Conniff, *The Energy People: A History of PSEG*. Public Service Electric and Gas Company, 1978.

Cooper, Carolyn. *Air Conditioning America: Engineers and the Controlled Environment, 1900–1960*. Johns Hopkins University Press, 1998.

Corn, Joseph. *The Winged Gospel*. Johns Hopkins University Press, 2002.

Cottrell, William Fred. *Energy and Society: The Relation Between Energy, Social Change, and Economic Development*. Greenwood, 1970.

Cowan, Ruth Schwartz. *More Work for Mother*. Basic Books, 1983.

Curvin, Robert, and Bruce Porter. *Blackout Looting, New York City, July 13, 1977*. Gardner, 1979.

Debard, Davis M. "What can American utilities learn from British war experience?" *Public Utilities Fortnightly* 29 (1942), no. 9: 531–539.

DeBeech, A. V. "Protection of electric public utility systems against air raids." *Edison Electric Institute Bulletin*, May 1942: 183–189.

de la Pena, Carolyn. *The Body Electric*. New York University Press, 2003.

Diamond, Jared. *Collapse: How Societies Choose to Fail or Succeed*. Penguin, 2005.

Dickerson, A. F. "Blackouts and dimouts in the United States." *American City* 57 (1942), October: 275–276.

Doheny-Farina, Stephen. *The Grid and the Village: Losing Electricity, Finding Community, Surviving Disaster*. Yale University Press, 2001.

Douglas, Ann. *Terrible Honesty: Mongrel Manhattan in the 1920s*. Farrar, Straus, and Giroux, 1996.

Douhet, Giulio. *The Command of the Air*. Arno, 1972.

Edison, Thomas A. (interview). "The woman of the future." *Good Housekeeping Magazine*, October 1913: 436.

Eggink, John. *Managing Energy Costs: A Behavioral and Non-Technical Approach*. Taylor & Francis, 2007.

Ehrlich, Paul. *The Population Bomb*. Ballantine Books, 1968.

Ekirch, A. Roger. *At Day's Close: A History of Nighttime*. Norton, 2005.

Energy Information Administration, *The Changing Structure of the Electric Power Industry: An Update*. US Department of Energy, 1996.

Fairley, Peter. "The unruly grid: One year later." *IEEE Spectrum*, August 2004: 22–27.

Fama, James. "Lessons of the blackout." *Mechanical Engineering* 126, no. 8 (2004): 36.

Farrell, Alexander E., Lester B. Lave, and Granger Morgan, "Bolstering the security of the electric power system." *Issues in Science and Technology*, spring 2002: 49–56.

Federal Energy Regulatory Commission. *The Con Edison Power Failure of July 13 and 14, 1977*. US Department of Energy, 1978.

Federal Power Commission. *Northeast Power Failure*. US Government Printing Office, 1965.

Federal Power Commission. *Prevention of Power Failures: An Analysis and Recommendations Pertaining to the Northeast Failure and the Reliability of US Power Systems*, volume 1. US Government Printing Office, 1967.

Foucault, Michel. "Des espace autres." *Architecture/Mouvement/Continuité*, October 1984.

Fox, Loren. *Enron: The Rise and Fall*. Wiley, 2003.

Friedlander, Gordon D. "What went wrong VIII: The Great Blackout of '65." *IEEE Spectrum*, October 1976: 84–85.

Friedrich, Jörn. *The Fire: The Bombing of Germany, 1940–1945*. Columbia University Press, 2006.

Frisch, Max. *Homo Faber*. Harcourt Brace, 1987.

Funigiello, Philip J. "Kilowatts for defense: The New Deal and the coming of the Second World War." *Journal of American History* 56 (1969), no. 3: 604–620.

Gerber, Michele Stenehjem. *On the Home Front: The Cold War Legacy of the Hanford Nuclear Site*, third edition. University of Nebraska Press, 2007.

Glick, Patricia, "The toll from coal: Power plants, emissions, wildlife and human health." *Bulletin of Science Technology Society* 21 (2001), no. 6: 482–500.

Goldberg, Arthur J. *Technological Change and Productivity in the Bituminous Coal Industry, 1920–1960.* Bulletin 1305, US Department of Labor, 1961.

Goodman, James. *Blackout.* North Point, 2003.

Graham, Stephen, and Nigel Thrift. "Out of order: Understanding repair and maintenance." *Theory, Culture, and Society* 24 (2007), no. 3: 1–25.

Grossman, Peter Z., and Daniel H. Cole, eds. *The End of a Natural Monopoly: Deregulation and Competition in the Electric Power Industry.* Routledge, 2003.

Halley, Arthur. *Overload.* Doubleday, 1979.

Hartzell, Karl Drew. *The Empire State at War.* State of New York, 1949.

Hibben, S. G. "Light and darkness in national defense." *Edison Electric Institute Bulletin*, July 1942: 261–263, 268.

Hilgartner, Steven, Richard C. Bell, Rory O'Connor. *Nukespeak: The Selling of Nuclear Technology in America.* Penguin, 1983.

Hirsh, Richard. *Power Loss: The Origins of Deregulation and Restructuring in the American Electric Utility System.* MIT Press, 1999.

Hirsh, Richard. *Technology and Transformation.* Cambridge University Press, 2003.

Hoffman, Bruce. *When the Lights Go Off and Never Come Back On: Nuclear Terrorism in America.* DIANE, 1987.

Hoffman, Steven M., and Angela High-Pippert. "Community energy: A social architecture for an alternative energy future." *Bulletin of Science, Technology, and Society* 25 (2005), no. 5: 387–401.

Howard, Luke. *Climate of London Deduced from Meteorological Observations.* Harvey and Darton, 1833.

Howe, John B. "A year after the blackout: On a collision course with history?" *Public Utilities Fortnightly* 142 (2004), no. 9: 18.

Hughes, Thomas P. "Technological momentum: Hydrogenation in Germany, 1900–1933." *Past and Present*, August 1969: 106–132.

Hughes, Thomas P. *Networks of Power*. Johns Hopkins University Press, 1983.

Hughes, Thomas P. "Technological momentum." In *Does Technology Drive History?* ed. Merritt Roe Smith and Leo Marx. MIT Press, 1994.

International Energy Agency. *Things That Go Blip in the Night: Standby Power and How to Limit It*. OECD, 2001.

Israel, Paul. *Edison: A Life of Invention*. Wiley, 1998.

Jackson, J. B. *Discovering the Vernacular Landscape*. Yale University Press, 1984.

Jakle, John A. *City Lights: Illuminating the American Night*. Johns Hopkins University Press, 2001.

Jameson, Robert, ed. *The US Power Market: Restructuring and Risk Management*. Risk Publications, 1997.

Jehl, Francis. *Menlo Park Reminiscences*. Edison Institute, 1937.

Kant, Immanuel. *Critique of Judgement*. Oxford University Press [1790], 1952.

Keating, Paul W. *Lamps for a Brighter America*. McGraw-Hill, 1954.

Keller, E. N. "How British utility accountants met the Blitz." *Edison Electric Institute Bulletin*, May 1942: 190–192.

Kennedy, David. *Freedom from Fear*. Oxford University Press, 1999.

Kenney, Dave. *Minnesota Goes to War: The Home Front During World War II*. Minnesota Historical Society, 2005.

Kirschen, Daniel, and Goran Strbac. "Why investments do not prevent blackouts." *Electricity Journal* 17 (2004), no. 2: 29–36.

Knell, Hermann. *To Destroy a City: Strategic Bombing and Its Human Consequences in World War II*. Da Capo, 2003.

Konvitz, Josef. "Why cities don't die." *American Heritage of Invention and Technology* 5 (1990), no. 3: 58–63.

LaCommare, Kristina Hamachi, and Joseph H. Eto, Ernest Orlando. *Understanding the Cost of Power Interruptions to US Electricity Consumers.* Report for the US Department of Energy, 2004.

Laitner, John A., and Karen Ehrhardt-Martinez. *Information and Communication Technologies: The Power of Productivity.* American Council for an Energy Efficient Economy, 2008.

Lambert, Jeremiah D. *Energy Companies and Market Reform: How Deregulation Went Wrong.* PennWell, 2006.

Landsberg, Helmut E. *The Urban Climate.* Academic, 1981.

Lara, Fernando. "In the dark all you have left is architecture." *Space and Culture* 9, no. 1 (2006): 26–27.

Lawrence Livermore National Laboratory. *The Jericho Option: Al-Qa'ida and Attacks on Critical Infrastructure.* UCRL-SR-224072, 2006.

Le Corbusier. *The Radiant City: Elements of a Doctrine of Urbanism to Be Used as the Basis of Our Machine-Age Civilization.* Orion, 1964.

Lesieutre, Bernard C., and Joseph H. Eto. *Electricity Transmission Congestion Costs: A Review of Recent Reports.* Report for US Department of Energy, 2003.

Lively, Kit. "California colleges struggle with blackouts." *Chronicle of Higher Education* 47 (2001), February 2: A27–A28.

Lopez, Barry. *Crossing Open Ground.* Vintage, 1989.

Lorant, Stefan. *Pittsburgh: The Story of an American City.* Doubleday, 1964.

Lovins, Amory B., and L. Hunter Lovins. *Brittle Power: Energy Strategy for National Security.* Brick House, 1982.

Lurkis, Alexander. *The Power Brink.* Icare, 1982.

Mahler, Jonathan. *The Bronx Is Burning: 1977, Baseball, Politics, and the Battle for the Soul of a City*. Farrar, Straus, and Giroux, 2005.

Main, Jeremy. "A peak load of trouble for the utilities." *Fortune*, November 1969: 118.

Makansi, Jason. *Lights Out: The Electricity Crisis, the Global Economy, and What It Means to You*. Wiley, 2007.

Makarov, Yuri V., Viktor I. Reshetov, Vladimir A. Stroev, and Nikolai I. Voropai. "Blackout prevention in the United States, Europe, and Russia." *Proceedings of the IEEE* 93 (2005), no. 11: 1942–1955.

Mamet, David. *Power Outage*. *New York Times*, August 6, 1977. Reprinted in *Goldberg Street: Short Plays and Monologues* (Grove, 1985).

Marling, Karal Ann. *As Seen on TV: The Visual Culture of Everyday Life in the 1950s*. Harvard University Press, 1994.

Marvin, Carolyn. *When Old Technologies Were New*. Oxford University Press, 1988.

Matthai, Karal Ann. *An Economic History of Women in America*. Schocken, 1982.

Maxwell, Alexander. "More about blackouts: Dissimilarities of British and American needs." *Edison Electric Institute Bulletin*, March 1942: 369–370.

Mazur, Allan. *A Hazardous Inquiry: The Rashomon Effect at Love Canal*. Harvard University Press, 1998.

McCraw, Thomas K. "Triumph and irony—the TVA." *Institute of Electrical and Electronic Engineers Proceedings* 64 (1976), September: 1372–1380.

McKinsey, Elizabeth. *Niagara Falls: Icon of the American Sublime*. Cambridge University Press, 1985.

McPhee, John. *The Curve of Binding Energy*. Farrar, Straus, and Giroux, 1974.

Meadows, Donella, et al. *The Limits to Growth*. Universe Books, 1972.

Moore, Francis C. *How to Build a Home: The House Practical*. Doubleday & McClure, 1897.

Moore, Fred. "High stakes, high voltage." *Computer Technology Review* 23 (2003), no. 9: 1.

Motter, Adilson E., and Ying-Cheng Lai. "Cascade-based attacks on complex networks." *Physical Review E* 66 (2002), 065102.

Mumford, Lewis. *The Myth of the Machine: The Pentagon of Power*. Harcourt Brace Jovanovich, 1964.

Munson, Richard. *From Edison to Enron: The Business of Power and What It Means for the Future of Electricity*. Praeger, 2005.

Nadar, Laura. "Barriers to thinking new about energy." *Physics Today* 9 (1981), February: 99–104.

Nadar, Laura. "The harder path—shifting gears." *Anthropological Quarterly* 77 (2004), no. 4: 771–791.

Nagel, Theodore J. "Operating a major electric utility today." *Science* 201 (1978), September 15: 985–993.

Nasaw, David. *Going Out*. Harvard University Press, 1999.

National Opinion Research Center. *Public Response to the Northeastern Power Blackout*. Report (1966) commissioned by US Office of Civil Defense.

Nidcic, Dusko P., Ian Dobson, Daniel S. Kirshen, Benjamin A. Carreras, and Vickie E. Lynch. "Criticality in a cascading failure blackout model." *International Journal of Electrical Power and Energy Systems* 28 (2006), no. 9: 627–633.

Nye, David E. *The Invented Self: An Anti-biography of Thomas Alva Edison*. University Press of Southern Denmark, 1983.

Nye, David E. *Image Worlds: Corporate Identities at General Electric*. MIT Press, 1985.

Nye, David E. *Electrifying America: Social Meanings of a New Technology*. MIT Press, 1990.

Nye, David E. *American Technological Sublime*. MIT Press, 1994.

Nye, David E. *Consuming Power: A Social History of American Energies*. MIT Press, 1998.

Nye, David E. "Electrifying expositions, 1880–1939." In Nye, *Narratives and Spaces: Technology and the Construction of American Culture*. Columbia University Press, 1998.

Nye, David E. "Path insistence: Comparing European and American attitudes toward energy." *Journal of International Affairs* 53 (1999), no. 1: 129–148.

Nye, David E. *America as Second Creation*. MIT Press, 2003.

Nye, David E. "From utopia to 'real-topia'—inventing the inevitable." In *Dreams of Paradise, Visions of Apocalypse*, ed. Jaap Verheul. VU University Press, 2004.

Nye, David E. "Are blackouts landscapes?" *American Studies in Scandinavia* 39 (2007), no. 2: 72–84.

Patterson, James. *Restless Giant: The United States from Watergate to Bush vs. Gore*. Oxford University Press, 2005.

Perrow, Charles, *Normal Accidents*. Basic Books, 1984.

Podair, Jarald. "Lights out." *Reviews in American History* 32 (2004), 267–273.

Ponting, Clive. *1940: Myth and Reality*. Hamish Hamilton, 1990.

Porter, Roy. *London: A Social History*. Harvard University Press, 1994.

Primeaux, W. J. *Direct Electric Utility Competition: The Natural Monopoly Myth*. Praeger, 1986.

Rifkin, Jeremy. *Entropy: A New World View*. Bantam, 1981.

Robinson, Charles Mulford. *Modern Civic Art*. Putnam, 1901.

Rocks, Lawrence, and Richard P. Runyon. *The Energy Crisis.* Crown, 1972.

Rose, Mark. *Cities of Light and Heat: Domesticating Gas and Electricity in Urban America.* Pennsylvania State University Press, 1995.

Rosenberg, Robert, Paul B. Israel, Keith A. Nier, and Melodie Andrews, eds. 1995. *Menlo Park: The Early Years, April 1876–December 1877. The Papers of Thomas A. Edison,* volume 3. Johns Hopkins University Press, 1995.

Roush, Wade Edmund. Catastrophe and Control: How Technological Disasters Enhance Democracy. Ph.D. dissertation, Massachusetts Institute of Technology, 1994.

Russell, I. Willis. "Among the new words." *American Speech* 20 (1945), no. 2: 141–146.

Savitz, David. "Case-control study of childhood cancer and exposure to 60-Hz magnetic fields." *American Journal of Epidemiology* 128 (1988): 21–38.

Scheer, Hermann. *Energy Autonomy: The Economic, Social and Technological Case for Renewable Energy.* Earthscan, 2007.

Schell, Bernadette H., and John L. Dodge. *The Hacking of America: Who's Doing It, Why, and How.* Quorum Books, 2002.

Schewe, Phillip F. *The Grid: A Journey Through the Heart of Our Electrified World.* Joseph Henry, 2007.

Schivelbusch, Wolfgang. "The force." *Cabinet* 21 (2006), spring.

Sears, John F. *Sacred Places: American Tourist Attractions in the Nineteenth Century.* Oxford University Press, 1989.

Sharpe, William Chapman. *New York Nocturne: The City After Dark in Literature, Painting, and Photography.* Princeton University Press, 2008.

Sherry, Michael. *The Rise of American Air Power: The Creation of Armageddon.* Yale University Press, 1987.

Silvestri, Mario. "Gli sviluppi technologici." In *Storia dell'indistria elettrica in Italia, 1926–1945*, volume 3, ed. Giuseppe Galasso. Laterza, 1993.

Singer, Jeremy. "Experts say United States is unpreprared for EMP attack." *Space News*, August 24, 2004.

Slocum, Tyson. "Electric utility deregulation and the myth of the energy crisis." *Bulletin of Science Technology Society* 21 (2001), no. 6: 473–481.

Smil, Vaclav. *Energies: An Illustrated Guide to the Biosphere and Civilization*. MIT Press, 1999.

Smil, Vaclav. *Energy at the Crossroads*. MIT Press, 2003.

Smil, Vaclav. *Creating the Twentieth Century: Technical Innovations of 1867–1914 and Their Lasting Impact*. Oxford University Press, 2005.

Smil, Vaclav. "The next 50 years: Fatal discontinuities." *Population and Development Review* 31 (2005), no. 2: 201–236.

Sovacool, Benjamin K. *The Dirty Energy Dilemma: What's Blocking Clean Power in the United States*. Praeger, 2008.

Stamp, L. Dudley. "Britain's coal crisis." *Geographical Review* 38 (1948), no. 2: 179–193.

Star, Susan Leigh. "The Ethnography of Infrastructure." *American Behavioral Scientist* 43 (1999), no. 3: 377–391.

Stern, Rudi. *The New Let There Be Neon*. ST Media Group International, 1996.

Swartz, Mimi, and Sherron Watkins. *Power Failure*. Doubleday, 2003.

Thomas, Donald. *An Underworld at War: Spivs, Deserters, Racketeers and Civilians in the Second World War*. John Murray, 2002.

Thomas, Robert J. "Managing relationships between electric power industry restructuring and grid reliability." Issue Papers on Reliability and Competition, 2005.

Thompson, Robert S. "'The air-conditioning capital of the world': Houston and climate control." In *Energy Metropolis*, ed. Martin Melosi and Joseph Pratt. University of Pittsburgh Press, 2007.

Thoreau, Henry David. *Walden*. Holt, Rinehart and Winston, 1948.

Turner, B. A. *Man-Made Disasters*. Wykeham Science Press, 1978.

Turner, Victor. *The Ritual Process: Structure and Anti-Structure*. Cornell University Press, 1969.

Turner, Victor. *Dramas, Fields and Metaphors: Symbolic Action in Human Society*. Cornell University Press, 1974.

Turner, Victor. "Frame, flow and reflection: Ritual and drama as public liminality." *Japanese Journal of Religious Studies* 6 (1979), no. 4: 465–499.

Tuttle, William M. "Pearl Harbor and America's homefront children: First fears, blackouts, air raid drills, and nightmares." Presented at Annual Meeting of American Historical Association, Pacific Coast Branch, 1991.

Tuttle, William M. *Daddy's Gone to War*. Oxford University Press, 1993.

Udry, J. Richard. "The effect of the Great Blackout of 1965 on births in New York City." *Demography* 7 (1970), no. 3: 325–327.

US Census Bureau. *Statistical Abstract of the United States*. US Government Printing Office, 2008.

US Department of Energy, Federal Energy Regulatory Commission. *The Con Edison Power Failure of July 13 and 14, 1977*. US Government Printing Office, 1978.

Van Dusen, Albert E. *Connecticut*. Random House, 1961.

Wainwright, Nicholas B. *History of the Philadelphia Electric Company, 1881–1961*. Philadelphia Electric Company, 1961.

Walker, J. Samuel. "Regulating against nuclear terrorism: The domestic safeguards issue, 1970–1979." *Technology and Culture* 42 (2001), no. 1: 107–132.

Ward, James. *Railroads and the Character of America, 1820–1887*. University of Tennessee Press, 1986.

Wattenberg, Ben. *The Statistical History of the United States*. Basic Books, 1976.

Weaver, Glen. *The Hartford Electric Light Company*. Helco, 1969.

Weber, Michael P. *Don't Call Me Boss: David L. Lawrence, Pittsburgh's Renaissance Mayor*. University of Pittsburgh Press, 1988.

Weil, Gordon L. *Blackout: How the Electric Industry Exploits America*. Nation Books, 2006.

Wertheimer, Nancy, and Ed Leeper. "Electrical wiring configurations and childhood cancer." *American Journal of Epidemiology* 109 (1979): 273–284.

White, Leslie A. *The Science of Culture*. Grove, 1949.

Wohlenberg, Ernest H. "The 'geography of civility' revisited: New York blackout looting, 1977." *Economic Geography* 58, no. 1: 42–48.

Wolak, Frank A. "Lessons from the California electricity crisis." In *Electricity Deregulation*, ed. James M. Griffin and Steven L. Puller. University of Chicago Press, 2005.

Wood, L. A. S. "Pedestrians should be seen and not hurt." *American City* 52 (1937), November: 32–34.

Yuill, Chris. "Emotions after dark—A sociological impression of the 2003 blackout." *Sociological Review Online* 9 (2004), no. 3.

Zachmann, Karin. "A socialist consumption junction: Debating the mechanization of housework in East Germany, 1956–1957." *Technology and Culture* 43 (2002), no. 1: 73–99.

Zanini, Michele, and Sean J. A. Edwards. "The networking of terror in the information age." In *Networks and Netwars*, ed. John Arquilla and David Ronfeldt. RAND Corporation, 2001.

Zweibel, Ken, James Mason, and Vasilis Fthenakis. "A solar grand plan." *Scientific American*, January 2008: 64–73.

Index

Abbey, Edward, 188
Accidents
 airline, 154, 155
 blackouts and, 6, 16, 33, 66, 71,
  84, 101–103, 160, 222
 causes of, 4, 16, 27, 29, 56, 69,
  84, 103, 112, 118, 120, 170,
  202
 "normal," 32, 103, 208
 Three Mile Island, 146, 199
Aesthetics, 5, 10–13, 98, 99,
  228–230
Air conditioning, 34, 79,
  109–114, 132, 133, 134, 143,
  165, 186, 213, 221, 225
al-Qaeda, 175, 177, 188, 194,
  202
Alternating current, 16, 17, 21
Anti-landscape, 130–132, 182,
  203, 231, 232
Apartments, 109, 111, 122, 123,
  134
Appliances, 16–20, 31, 74, 78, 79,
  109, 112, 116, 118, 124, 127,

143–145, 200, 202, 212, 213,
  225, 231
Architecture, 15, 73, 74, 91, 98,
  110, 111, 184, 210, 214
Army Corps of Engineers, 27
Arson, 105, 106, 123, 124
Atlanta, 14, 15, 23, 220, 221
Aviation, 2, 37, 40–55, 75, 89,
  102, 154, 155, 186, 187, 193,
  230

Baghdad, 199, 200
Bakhtin, 183
Batteries, 76, 77, 121
Berlin, 44
Beston, Henry, 9, 10
Blackout (film), 134
Blackouts, 2–4, 39, 67–69, 77, 80,
  206, 207
 1936, 71, 72, 80, 81, 84, 101, 13
 1965, 2, 3, 28, 69–72, 81–95, 99,
  101–107, 118, 120, 127, 132,
  133, 183, 190, 208
 1977, 3, 81, 101, 105–136, 185

Blackouts (cont.)
  2003, 4, 32, 33, 81, 101, 155,
    156, 160–164, 174–180,
    183–186, 192, 229
  and birth rate, 94
  crime during, 3, 52, 89, 90, 105,
    123–128, 133–135, 181, 182,
    185, 203, 208
  drills for, 40–48
  economic effects of, 19, 29,
    59–64, 78, 185, 186, 205,
    206, 231
  EMP and, 200–203
  frequency of, 3, 17, 20, 27–29,
    139, 148, 149, 153, 166–171,
    221, 225
  improvisation during, 33, 34,
    76, 81, 82, 86, 90, 96, 181–184,
    190, 191, 195, 203, 208, 209,
    225
  invented, 13, 15, 19, 20
  memories of, 1, 2, 72, 94,
    98, 99
  military, 2, 5, 37–60, 69, 70,
    206, 207
  modeled, 167–171
  rolling, 3, 136–138, 144–154,
    165, 166, 207
  sound of, 5, 83, 96, 229
  strikes and, 59–64
Bombing, 2, 40–51, 55, 65, 186,
    187
Boston, 62, 86, 93, 103
Britain, 47–50, 55–58, 64, 65,
    135, 167, 173, 217, 220, 221
Broadway, 10, 11, 56, 57, 59, 63,
    121, 124, 134, 206

Brownouts
  strikes and, 62, 63
  utility, 108, 136–138, 144, 198,
    210, 216, 224
  in war, 56, 57
Buenos Aires, 138
Bureau of Reclamation, 27
Bush, George H. W., 141
Bush, George W., 216

California, 28, 117, 132,
    147–153, 162, 191, 227
Callenbach, Ernest, 188
Canada, 4, 24, 69, 84, 166, 185,
    190, 219, 220
Cancer, 159
Candles, 15, 60, 72, 74, 86,
    91–94, 230, 231
Carbon dioxide, 116, 217,
    225–227, 231
Carter, Jimmy, 116, 117, 139
Central Intelligence Agency,
    196
Chicago, 14, 20, 21, 61, 62, 96,
    148
Children, 48, 53, 54, 58, 74, 76,
    81, 114, 135, 159
China, 225
Clarke, Lee, 6, 180, 181
Clinton, Bill, 141, 179, 180
Coal, 24, 28, 30, 31, 56, 61–65,
    106–108, 114–117, 146, 219,
    226, 227
Colburn, Alvin Langdon, 98
Cold War, 78, 79
Columbian Exposition (1893),
    20, 21, 96

Computers, 67, 70, 76, 79, 81, 151, 153, 162, 184, 186, 194, 196, 225

Connecticut, 42

Consolidated Edison, 59, 68, 71, 72, 75, 84, 85, 97, 107, 117–120, 135

Dairy industry, 152, 153

Dark Sky Association, 229

Defamiliarization, 9, 10, 34, 35, 40, 98

Department of Defense, 27, 195

Department of Energy, 25, 32, 162, 153, 162, 165, 210

Department of Homeland Security, 3, 195

Department stores, 71, 72, 93, 97

Deregulation
  of airlines, 154, 155
  of electrical system, 6, 20, 138–143, 147–154, 159, 162, 163, 224
  of telephone system, 155–157

Direct current, 17, 20, 21

Disaster planning, 28, 179–181, 209

Doherty, Henry, 24, 25

Douhet, Giulio, 42, 43

Doyle, John, 6, 167–170

Dresden, 55

Dublin, 220

Earth Hour, 216–220

Economies of scale, 21–27, 114, 115, 138–142. *See also* Energy efficiency

Eco-terrorism, 188

Edison, Thomas, 13, 14, 17, 19, 21 97, 205, 206, 223

Ekirch, A. Roger, 14

Electrical medicine, 75, 76

Electrical metaphors, 76, 77

Electric belts, 75, 76

Electricity
  consumption of, 2, 3, 18–27, 30, 48, 64, 74, 75, 78, 107–109, 116, 137, 143, 144, 210, 213, 214, 216, 228
  dependence on, 3, 4, 27, 59–61, 68–73, 78, 81, 85–90, 108, 129–132, 178, 181, 182, 185, 186, 200, 205–207, 224
  naturalization of, 10–13, 19, 34, 35, 58, 67, 79, 80, 98, 101, 132, 170, 206, 222, 223
  production of, 15–21, 24–29, 48, 49, 56, 64, 65, 137–146, 163, 164, 210, 227, 231
  rates, 5, 22, 23, 107, 138–140, 147–151, 223

Electric Power Research Institute, 29, 159

Electromagnetic pulse, 200–203

Elevators, 68, 71, 82, 86, 92

Energy Crisis, 3, 107–117, 132, 136, 139, 144

Energy efficiency, 114, 115, 138, 140, 142–147, 198, 207, 210–214, 228, 231, 232

Energy intensity, 212–214

Energy as measure of civilization, 77, 78

Energy Policy At of 1992, 142

Energy transition, 64, 65, 227
Energy Transmission Act of 2005, 164, 165, 205
Enron Corporation, 148–151, 154, 158, 224
Escape from New York (film), 134
Europe, 29, 30, 135, 166, 167, 213, 220, 226

Federal Bureau of Investigation, 174
Federal Emergency Management Administration, 182
Federal Energy Regulatory Commission, 158, 165
Federal Power Commission, 25
Ferlinghetti, Lawrence, 34
Fire departments, 70, 90, 105, 106, 108, 121, 123, 178
Florida, 137, 220, 221
Ford Motor Company, 62
Fort Bragg, North Carolina, 46, 47
Foucault, Michel, 6, 95, 96, 129, 209
Fox News, 219
France, 44, 59, 64, 146
Frisch, Max, 34, 35

Gas turbines, 140, 146, 150
General Electric, 22, 79, 97, 114
Germany, 40, 44, 45, 48, 55, 211, 226, 227
Gibraltar, 41
Global positioning systems, 66
Global warming, 3, 113, 225, 231

"Gonna Be a Blackout Tonight" (song), 53
Gore, Al, 227
Grand Canyon, 9, 12
Grand Coulee Dam, 26, 28
Great White Way, 10, 11, 206
Green Building Council, 214
Greenouts, 3, 216–222, 229, 232
Grid
  and blackouts, 68–71, 75, 85, 87, 102, 103, 109, 114, 118–120, 132–135
  construction of, 2, 13–35
  overloaded, 156–159, 167–171, 226
Gutman, Herbert, 128

Hartford, 16, 45, 85
Hausenstein, Wilhelm, 229, 230
Heat islands, 111–113
Heidegger, Martin, 34
Heterotopia, 6, 95, 96, 131, 209
Hirsh, Richard, 5, 115, 139, 140
Hitchcock, Alfred, 182
Home electric generators, 197, 198, 225
Hoover Dam, 25, 28
Hoover, Herbert, 205, 206
Hospitals, 50, 86, 102, 105, 122, 123, 129, 132, 138
Houston, 111–113, 137
Hughes, Thomas P., 5, 16, 17, 232
Hurricanes, 49, 181, 182, 198
Hydroelectric dams, 24, 25, 26, 28, 61, 84, 85, 97, 138, 146, 147, 148, 188, 196, 225

Ice storms, 103, 190

"I'm Gonna Git Lit Up When the Lights Go Up in London" (song), 57

Incessance, 99, 100

India, 28, 225

Insull, Samuel, 21, 22

Internet, 156, 184, 186, 194–196, 203

Iraq, 186, 199, 200, 216

Irish Republican Army, 173

Istanbul, 41

Italy, 29

Jackson, J. B., 6, 130

Japan, 41, 42, 63, 64

Kansas City, 59, 60

Kant, Immanuel, 12

Khrushchev, Nikita, 78

Kitchens, 18, 78, 79, 143.
  See also Appliances

Kyl, Jon, 202, 203

Labor conflicts, 33, 59–64

Landscape. *See also* Anti-landscape; Sublime; Technological Sublime
  blackout as, 40, 46, 47, 96–99, 128–131, 134, 184
  electric, 6, 9, 10, 58, 66, 75, 229

Language and electricity, 76, 77

Las Vegas, 9, 10, 80, 110

Le Corbusier, 59

Lighting
  domestic, 14, 15, 18, 23, 64, 74, 79, 143, 213, 230
  emergency, 70
  fluorescent, 111
  gas, 10, 15, 21, 59, 133
  urban, 10–15, 20, 35, 62, 79, 80, 95, 112, 133, 205, 213, 216–220, 223, 229
  World War II and, 39, 50, 56–59, 62

Lightning, 16, 27, 69, 84, 103, 112, 118, 120, 170, 202

Light pollution, 9, 229

Liminal space, 9, 82, 83, 92, 94, 95, 101, 105, 129, 130, 132, 184, 208, 209

Limits to Growth, The (book), 116

Live Free or Die Hard (film), 192, 195

Load
  balancing of, 17–24, 29, 31, 102, 103, 111, 115, 142, 161, 162
  lost, 28, 155
  shedding of, 84, 102, 118, 119, 144, 145, 192

Lockheed Electronics, 145

London, 12, 21, 43, 44, 50–52, 58, 66, 133, 134, 167, 173, 181–183, 217

Looters, 105, 118, 123–135

Los Angeles, 7, 12, 25, 52, 53, 111–113, 165, 189

Love Canal, 131

Lovins, Amory and Hunter, 132

Malta, 41

"Miami 2017" (song), 134

Mississippi Power and Light, Co, 24

Morton, Rogers, 108

Mumford, Lewis, 68

Murrow, Edward R., 50, 51

Nadar, Laura, 144

Natural gas, 10, 15, 21, 27, 29, 30, 31, 65, 107, 108, 140, 146, 150, 210

Natural monopoly, 21–23, 120, 138–142, 223, 224

Newark, 72

New England, 85, 137, 165

New Jersey, 16, 17, 60, 72, 85, 158, 165, 166

New Orleans, 181, 182, 203

New York City

blackouts in, 28, 33, 59, 68–71, 81–98, 101, 105–107, 117–129, 133–135, 160, 165, 183–185, 190, 207, 208, 221, 229

electrification of 10–12, 21, 31, 37, 38, 61, 62, 78, 205–208, 223

terrorism and, 174, 175, 180, 189

World War II and, 6, 42, 45, 53–58, 61–63

New York State, 24, 84, 85, 165, 175, 190, 221

Niagara Falls, 24, 28, 84, 85, 97

Night, experience of, 9, 10, 13, 14, 67, 206

Nixon, Richard, 78, 108, 141

Nocturne, 98

North American Electric Reliability Corporation, 102, 103, 157, 160–162, 166, 169, 180

Nuclear weapons, 179, 200–203

Obama, Barack, 214, 227

Off-grid living, 210, 212

Office buildings, 72, 73, 91, 109, 111, 112

Oil, 65, 106–108, 114, 117, 146, 164, 180, 185, 188, 199, 212

Overload (novel), 191

Pacific Gas and Electric, 136, 145, 147, 148, 213, 218

Paris, 59, 64

Patriot Games (novel), 191

Perrow, Charles, 6, 32, 208

Philadelphia, 17, 31, 62

Phonograph, 19

Photography, 75, 98, 99

Pictorialism, 98, 99

Pinchot, Gifford, 24

Pittsburgh, 9, 60, 61

PJM system, 158, 164, 166

Police, 52, 70, 83, 88, 89, 95, 105, 123–126, 132, 134

Pollution, 9, 35, 113, 116, 131, 225, 229

Power. *See also* Power Plants

decentralized, 15–17, 27, 31, 132, 196–199, 210, 223, 224, 227, 228

solar, 29, 30, 31, 210–212, 226, 227

wind, 29, 30, 31, 140, 197, 210, 212, 213, 224–227, 231

Power Outage (play), 128
Power plants
  early, 15–17, 20–24
  efficiency of, 21, 114, 115, 145, 146
  gas turbine, 140, 146, 150, 198
  nuclear, 77, 78, 108, 117, 118, 121, 146, 174, 179, 199, 225, 226, 231
  steam-powered, 21, 22, 61, 106, 188, 226
  technical limits of, 114, 115, 145, 146
  water-powered, 227
Public Utility Regulatory Policies Act, 139, 140

Radio, 67, 76, 86, 87, 99, 110, 112, 116, 122, 132, 217
Railroads, 222, 223
Reagan, Ronald, 141
Regulation, 22, 23
Richardson, Bill, 160
Riots, 105, 118, 123–135, 185
Roosevelt, Franklin D., 25
Ross, Harold, 5
Roush, Wade, 132
Rural Electrification Administration, 25–27
Russia, 166

Sabotage (film), 182
San Francisco, 218, 220
Satellites, 67, 76
Searchlights, 43, 46, 47
Serbia, 186
Shopping malls, 73

Silicon Valley, 151
Simpsons, The (television series), 133, 134
Sky glow, 9, 35, 229
Skyline, 12, 13, 54, 98, 99, 177, 207, 218
Skyscrapers, 12, 37, 78, 89, 94, 96–99, 102, 111, 131, 213
Socialism, 24, 79
South Africa, 166
Star, Susan Leigh, 34
Statue of Liberty, 57, 58, 94, 95
Stieglitz, Alfred, 98, 99
Stranded costs, 150
Sublime, 10–13, 58. *See also* Technological sublime
Suburbia, 73, 112–114, 131
Subways, 68, 71, 82, 87, 88, 92, 96, 102, 121, 129, 217
Sydney, 220

Technological momentum, 5, 16, 17, 223, 227, 232
Technological sublime, 10–13, 19, 58, 59, 65, 66, 94, 95, 222, 232
Telephones, 33, 142, 155–157, 184, 193
Tennessee Valley Authority, 25, 28, 61
Terrorism, 173–203, 207
Thoreau, Henry David, 34
Thrift, Nigel, 19
Times Square, 12, 54, 57, 59, 61, 63, 66, 80, 121, 174, 205
Toronto, 69, 85, 219, 220

Transmission lines. *See also* Grid
and blackouts, 111, 118, 119, 135
early, 13–30
and rolling blackouts, 149, 150,
154–165, 168, 169
vulnerability of, 49, 50, 174,
175, 180, 186–193, 197
Truman, Harry, 62
Tucson, 229
Turner, Victor, 6, 82, 83, 92, 95,
129, 184

Unintended consequences, 10,
13, 28, 69, 141, 207
Utilities, 4, 21–31, 59–63, 107,
108, 114, 115, 158, 189, 190,
212, 223–225

War Production Board, 56
"When the Lights Go On Again
(All Over the World)" (song),
2, 56, 57
*Where Were You When the Lights
Went Out?* (film), 94
Whistler, James McNeill, 98
White, Leslie, 77, 78
Will, George F., 13
Wood, Lowell, 201, 202
World's Fairs, 14, 20, 21, 75,
96, 97
World War I, 40
World War II, 2, 26, 48–59
Wright, Frank Lloyd, 74

Y2K fears, 134